Vector Optimization

Editor:
Johannes Jahn
University of Erlangen-Nürnberg
Department of Mathematics
Martensstr. 3
91058 Erlangen
Germany
jahn@am.uni-erlangen.de

Vector Optimization

The series Vector Optimization contains publications in various fields of optimization with vector-valued objective functions, such as multiobjective optimization, multi criteria decision making, set optimization, vector-valued game theory and border areas to financial mathematics, biosystems, semidefinite programming and multiobjective control theory. Studies of continuous, discrete, combinatorial and stochastic multiobjective models in interesting fields of operations research are also included. The series covers mathematical theory, methods and applications in economics and engineering. These publications being written in English are primarily monographs and multiple author works containing current advances in these fields.

Gabriele Eichfelder

Adaptive Scalarization Methods in Multiobjective Optimization

Springer

Author:
Dr. Gabriele Eichfelder
University of Erlangen-Nürnberg
Department of Mathematics
Martensstr. 3
91058 Erlangen
Germany
Gabriele.Eichfelder@am.uni-erlangen.de

ISBN 978-3-540-79157-7   e-ISBN 978-3-540-79159-1

Library of Congress Control Number: 2008924782

© 2008 Springer-Verlag Berlin Heidelberg

This work is subject to copyright. All rights are reserved, whether the whole or part of the material is concerned, specifically the rights of translation, reprinting, reuse of illustrations, recitation, broadcasting, reproduction on microfilm or in any other way, and storage in data banks. Duplication of this publication or parts thereof is permitted only under the provisions of the German Copyright Law of September 9, 1965, in its current version, and permissions for use must always be obtained from Springer-Verlag.
Violations are liable for prosecution under the German Copyright Law.

The use of general descriptive names, registered names, trademarks, etc. in this publication does not imply, even in the absence of a specific statement, that such names are exempt from the relevant protective laws and regulations and therefore free for general use.

Cover design: WMXDesign GmbH, Heidelberg, Germany

Printed on acid-free paper

9 8 7 6 5 4 3 2 1

springer.com

To my family,
Paul, Sabine, Susanne, and Claudia,
and especially to Tom.

Preface

In many areas in engineering, economics and science new developments are only possible by the application of modern optimization methods. The optimization problems arising nowadays in applications are mostly multiobjective, i. e. many competing objectives are aspired all at once. These optimization problems with a vector-valued objective function have in opposition to scalar-valued problems generally not only one minimal solution but the solution set is very large. Thus the development of efficient numerical methods for special classes of multiobjective optimization problems is, due to the complexity of the solution set, of special interest. This relevance is pointed out in many recent publications in application areas such as medicine ([63, 118, 100, 143]), engineering ([112, 126, 133, 211, 224], references in [81]), environmental decision making ([137, 227]) or economics ([57, 65, 217, 234]).

Considering multiobjective optimization problems demands first the definition of minimality for such problems. A first minimality notion traces back to Edgeworth [59], 1881, and Pareto [180], 1896, using the natural ordering in the image space. A first mathematical consideration of this topic was done by Kuhn and Tucker [144] in 1951. Since that time multiobjective optimization became an active research field. Several books and survey papers have been published giving introductions to this topic, for instance [28, 60, 66, 76, 112, 124, 165, 188, 189, 190, 215]. In the last decades the main focus was on the development of interactive methods for determining one single solution in an iterative process. Thereby numerical calculations alternate with subjective decisions of the decision maker (d. m.) till a satisfying solution is found. For a survey of interactive methods see [28, 124, 165].

Based on an extreme increase in computer performances it is now possible to determine the entire efficient set. Having an approximation of the whole solution set available the decision maker gets a useful insight in the problem structure and important information are delivered like trade-off information. Thereby trade-off is the information how the improvement of one objective function leads to a deterioration of the other objectives. The importance of approximating the complete efficient set is thus also emphasized in many applications. Especially in engineering it is interesting to know all design alternatives ([119]). Hence nowadays there is an increasing interest in methods for approximating the whole solution set as also the high number of papers related to this topic demonstrates, see for instance [10, 40, 82, 81, 84, 83, 106, 139, 164, 182, 196, 197].

For the determination of approximations of the efficient set several approaches have been developed, as for example evolutionary algorithms (for surveys see [31, 41, 112, 228, 246]) or stochastic methods ([194]). A large class of methods is based on scalarizations. This means the replacement of the multiobjective optimization problem by a suitable scalar optimization problem involving possibly some parameters or additional constraints. Examples for such scalarizations are the weighted sum ([245]) or the ε-constraint problem ([98, 159]). In this book we concentrate on the scalarization approach and we set especially value on the scalar problem according to Pascoletti and Serafini ([181]). However, many other existing auxiliary problems, which will also be presented, can be related to that method.

As generally not the entire efficient set can be computed an approximation is instead generated by solving the scalar problems for various parameters. The information delivered to the decision maker by such an approximation depends mainly on the quality of the approximation. Too many points are related to a high computational effort. Too few points means that some parts of the efficient set are neglected. Hence it is important to take quality criteria as discussed for instance in [32, 43, 101, 141, 191] into account. An approximation with a high quality is given if it is stinted but also representative, i.e. if the approximation points are spread evenly over the efficient set with almost equal distances.

We develop in this book methods for generating such approximations for nonlinear differentiable problems. For these methods the sensitivity of the scalar problems on their parameters are examined. These sensitivity results are used for developing an adaptive parameter control. Then, without any interaction from the decision maker, the choice of the parameters is in such a way controlled during the procedure, that the generated approximation points have almost equal distances.

Thereby we consider very general multiobjective problems and allow arbitrary partial orderings induced by a closed pointed convex cone in the image space (like in [81, 96, 106, 181, 230]) using the notion of K-minimality as given in [14, 102, 122, 124, 190, 243]. The partial ordering of the Edgeworth-Pareto-minimality concept represented by the natural ordering cone, the positive orthant, is included as a special case. More general orderings rise the applicability of our methods as the decision makers get more freedom in the formulation of the optimization problems. Preference structures can be incorporated, which cannot be expressed explicitly by an objective function (see Example 1.5 and [230, Example 4.1]). In decision theory in economics it is a well-known tool to use arbitrary partial orderings for modeling the relative importance of several criteria for a d.m. as well for handling groups of decision makers ([235]).

For example in [116, 117] convex polyhedral cones are used for modeling the preferences of a d.m. based on trade-off information facilitating multi-criteria decision making. In portfolio optimization ([5]) polyhedral cones in $\mathbb{R}^m$ generated by more than m vectors, as well as non-finitely generated cones as the ice-cream cone, are considered. Besides, orderings, other than the natural ordering, are important in [85] where a scalar bilevel optimization problem is reformulated as a multiobjective problem. There a non-convex cone which is the union of two convex cones is used. Helbig constructs in [106] cone-variations as a tool for finding EP-minimal points, see also [134, 237]. In addition to that Wu considers in [238] convex cones for a solution concept in fuzzy multiobjective optimization. Hence, multiobjective optimization problems w.r.t. arbitrary partial orderings are essential in decision making and are further an important tool in other areas. Therefore we develop our results w.r.t. general partial orderings.

This book consists of three parts. In the first part theoretical basics of multiobjective optimization are introduced as for instance minimality notions and properties of ordering cones especially of polyhedral cones. Scalarizations are discussed with a special focus on the Pascoletti-Serafini scalarization. Further, sensitivity results for these parameter depended scalar problems are developed like the first order derivative information of the local minimal value function.

The second part is devoted to numerical methods and their application. Quality criteria for approximations of the efficient set are introduced and the main topic of this book, the adaptive parameter control using the sensitivity results developed before, is constructed. We differentiate thereby between the treatment of biobjective optimization problems and general multiobjective optimization problems. The gained algorithms are applied to various test problems and to an actual application in intensity modulated radiotherapy.

The book concludes in the third part with the examination of multiobjective bilevel problems and a solution method for those kinds of problems, which is also applied to a test problem and to an application in medical engineering.

I am very grateful to Prof. Dr. Johannes Jahn for his support as well as to Prof. Dr. Joydeep Dutta, Prof. Dr. Jörg Fliege and PD Dr. Karl-Heinz Küfer for valuable discussions. Moreover, I am indebted to Dipl.-Math. Annette Merkel, Dr. Michael Monz, Dipl.-Technomath. Joachim Prohaska and Elizabeth Rogers.

Erlangen, January 2008   *Gabriele Eichfelder*

Contents

Part I Theory

1 Theoretical Basics of Multiobjective Optimization ... 3
  1.1 Basic Concepts 3
  1.2 Polyhedral Ordering Cones 15

2 Scalarization Approaches 21
  2.1 Pascoletti-Serafini Scalarization 23
  2.2 Properties of the Pascoletti-Serafini Scalarization .. 25
  2.3 Parameter Set Restriction for the Pascoletti-Serafini
      Scalarization 31
      2.3.1 Bicriteria Case 32
      2.3.2 General Case 40
  2.4 Modified Pascoletti-Serafini Scalarization 44
  2.5 Relations Between Scalarizations 49
      2.5.1 ε-Constraint Problem 49
      2.5.2 Normal Boundary Intersection Problem 53
      2.5.3 Modified Polak Problem 55
      2.5.4 Weighted Chebyshev Norm Problem 57
      2.5.5 Problem According to Gourion and Luc 58
      2.5.6 Generalized Weighted Sum Problem 59
      2.5.7 Weighted Sum Problem 61
      2.5.8 Problem According to Kaliszewski 65
      2.5.9 Further Scalarizations 66

3 Sensitivity Results for the Scalarizations 67
  3.1 Sensitivity Results in Partially Ordered Spaces 68

XII   Contents

    3.2   Sensitivity Results in Naturally Ordered Spaces 83
    3.3   Sensitivity Results for the ε-Constraint Problem 94

Part II Numerical Methods and Results

4 Adaptive Parameter Control 101
    4.1   Quality Criteria for Approximations 101
    4.2   Adaptive Parameter Control in the Bicriteria Case 107
        4.2.1   Algorithm for the Pascoletti-Serafini Scalarization   123
        4.2.2   Algorithm for the ε-Constraint Scalarization   128
        4.2.3   Algorithm for the Normal Boundary Intersection
               Scalarization 130
        4.2.4   Algorithm for the Modified Polak Scalarization . . . 132
    4.3   Adaptive Parameter Control in the Multicriteria Case . . 134

5 Numerical Results 141
    5.1   Bicriteria Test Problems 141
        5.1.1   Test Problem 1: ε-Constraint Scalarization 141
        5.1.2   Test Problem 2: Comparison with the Weighted
               Sum Method 144
        5.1.3   Test Problem 3: Non-Convex Image Set 147
        5.1.4   Test Problem 4: Non-Connected Efficient Set 149
        5.1.5   Test Problem 5: Various Ordering Cones 153
    5.2   Tricriteria Test Problems 155
        5.2.1   Test Problem 6: Convex Image Set 155
        5.2.2   Test Problem 7: Non-Convex Image Set 158
        5.2.3   Test Problem 8: Comet Problem 161
        5.2.4   Test Problem 9: Non-Connected Efficient Set 163

6 Application to Intensity Modulated Radiotherapy 167
    6.1   Problem Formulation Using a Bicriteria Approach 168
    6.2   Problem Formulation Using a Tricriteria Approach 176

Part III Multiobjective Bilevel Optimization

7 Application to Multiobjective Bilevel Optimization . . 183
    7.1   Basic Concepts of Bilevel Optimization 184
    7.2   Induced Set Approximation 186
    7.3   Induced Set Refinement 193

7.4	Algorithm	197
7.5	Numerical Results	199
	7.5.1 Test Problem	199
	7.5.2 Application Problem	202
7.6	Multiobjective Bilevel Optimization Problems with Coupled Upper Level Constraints	210

References 219

Index 239

Part I

Theory

1
Theoretical Basics of Multiobjective Optimization

Multiobjective optimization is an indispensable tool for decision makers if the benefit of a decision does not depend on one objective only, which then can be mapped by one scalar-valued function, but if several competing objectives are aspired all at once. In this chapter we discuss basic concepts of vector optimization such as minimality notions based on partial orderings introduced by convex cones.

1.1 Basic Concepts

Our aim is a minimization of a vector-valued objective function $f\colon \mathbb{R}^n \to \mathbb{R}^m$ ($n, m \in \mathbb{N}$) subject to constraints. We consider optimization problems defined by

$$\text{(MOP)} \quad \min f(x)$$

subject to the constraints
$$g(x) \in C,$$
$$h(x) = 0_q,$$
$$x \in S$$

with given continuous functions $f\colon \mathbb{R}^n \to \mathbb{R}^m$, $g\colon \mathbb{R}^n \to \mathbb{R}^p$, and $h\colon \mathbb{R}^n \to \mathbb{R}^q$ ($p, q \in \mathbb{N}$). We set $f(x) = (f_1(x), \ldots, f_m(x))$ with $f_i\colon \mathbb{R}^m \to \mathbb{R}$, $i = 1, \ldots, m$. Let $S \subset \mathbb{R}^n$ be a closed convex set and $C \subset \mathbb{R}^p$ be a closed convex cone. A set $C \subset \mathbb{R}^p$ is a convex cone if $\lambda(x+y) \in C$ for all $\lambda \geq 0$, $x, y \in C$. For $m = 1$ the problem (MOP) reduces to a standard optimization problem in nonlinear optimization

with a scalar-valued objective function. However in multiobjective optimization we are interested in several competing objective functions and thus we assume $m \geq 2$. The set
$$\Omega := \{x \in S \mid g(x) \in C, \ h(x) = 0_q\}$$
is called the constraint set of the problem (MOP). We assume $\Omega \neq \emptyset$ and define
$$f(\Omega) := \{f(x) \in \mathbb{R}^m \mid x \in \Omega\}.$$
We will discuss only minimization problems. However each maximization problem can be transformed to a minimization problem very easily by considering the negative objective function values.

We are looking for the minimal values of the function f over the set Ω. In general there is not one solution satisfying all objectives best at the same time. We assume that the decision maker (d. m.) specifies which alternative x he prefers to another point x' by declaring if he prefers $f(x)$ to $f(x')$ or not. With these statements a binary relation is defined in the image space $\mathbb{R}^m$ (and in the parameter space $\mathbb{R}^n$ respectively), also called preference order (see [190, 230, 232]). Different types of preference orders are possible but in practical applications a very common concept are partial orderings ([150]). We recall the definition of a binary relation and of partial orderings in the Definition 1.1.

Another approach for specifying preferences of the d. m. are the so-called domination structures by Yu ([243], also discussed for instance in [152, 244]) where the preference order is represented by a set-valued map. There, for all $y \in \mathbb{R}^m$, the set
$$D(y) := \{d \in \mathbb{R}^m \mid y \succ y + d\} \cup \{0_m\}$$
is defined (see also [190, p.28]) where $y \succ y'$ means that the d. m. prefers y more than y'. A deviation of $d \in D(y)$ from y is hence less preferred than the original y. The most important and interesting special case of a domination structure is when $D(\cdot)$ is a constant set-valued map, especially if $D(y)$ equals a pointed convex cone for all $y \in \mathbb{R}^m$, i.e. $D(\cdot) = K$. A cone K is called pointed if $K \cap (-K) = \{0_m\}$. The special case of a polyhedral domination cone is considered in [177, 220]. For the definition of a polyhedral cone see Sect. 1.2. A variable preference model is for instance discussed in [71].

The concept of a constant set-valued map defining a domination structure has a direct connection to partial orderings. We recall the definition of partial orderings:

1.1 Basic Concepts

Definition 1.1. (a) A nonempty subset $R \subset \mathbb{R}^m \times \mathbb{R}^m$ is called binary relation R on $\mathbb{R}^m$. We write xRy for $(x, y) \in R$.
(b) A binary relation $\leq$ on $\mathbb{R}^m$ is called partial ordering on $\mathbb{R}^m$ if for arbitrary $w, x, y, z \in \mathbb{R}^m$:
  (i) $x \leq x$ (reflexivity),
  (ii) $x \leq y$, $y \leq z \Rightarrow x \leq z$ (transitivity),
  (iii) $x \leq y$, $w \leq z \Rightarrow x + w \leq y + z$ (compatibility with the addition),
  (iv) $x \leq y$, $\alpha \in \mathbb{R}_+ \Rightarrow \alpha x \leq \alpha y$ (compatibility with the scalar multiplication).
(c) A partial ordering $\leq$ on $\mathbb{R}^m$ is called antisymmetric if for arbitrary $x, y \in \mathbb{R}^m$
$$x \leq y, \ y \leq x \Rightarrow x = y.$$

A linear space $\mathbb{R}^m$ equipped with a partial ordering is called a partially ordered linear space. An example for a partial ordering on $\mathbb{R}^m$ is the natural (or componentwise) ordering $\leq_m$ defined by

$$\leq_m := \{(x, y) \in \mathbb{R}^m \times \mathbb{R}^m \mid x_i \leq y_i \text{ for all } i = 1, \ldots, m\}.$$

Partial orderings can be characterized by convex cones. Any partial ordering $\leq$ in $\mathbb{R}^m$ defines a convex cone by

$$K := \{x \in \mathbb{R}^m \mid 0_m \leq x\}$$

and any convex cone, then also called ordering cone, defines a partial ordering on $\mathbb{R}^m$ by

$$\leq_K := \{(x, y) \in \mathbb{R}^m \times \mathbb{R}^m \mid y - x \in K\}.$$

For example the ordering cone representing the natural ordering in $\mathbb{R}^m$ is the positive orthant $\mathbb{R}^m_+$. A partial ordering $\leq_K$ is antisymmetric if and only if K is pointed. Thus, preference orders which are partial orderings correspond to domination structures with a constant set-valued map being equal to a convex cone.

With the help of orderings introduced by ordering cones K in $\mathbb{R}^m$ we can define minimal elements of sets in $\mathbb{R}^m$ ([22, 102, 108, 122, 124]).

Definition 1.2. Let T be a nonempty subset of the linear space $\mathbb{R}^m$ partially ordered by the convex cone K. A point $\bar{y} \in T$ is a K-minimal point of the set T if

$$(\bar{y} - K) \cap T \subset \bar{y} + K. \tag{1.1}$$

1 Theoretical Basics

If the cone K is pointed then (1.1) is equivalent to

$$(\bar{y} - K) \cap T = \{\bar{y}\}.$$

If for $y, \tilde{y} \in T$ we have $y - \tilde{y} \in K \setminus \{0_m\}$, then we say $\tilde{y}$ dominates y (Fig. 1.1).

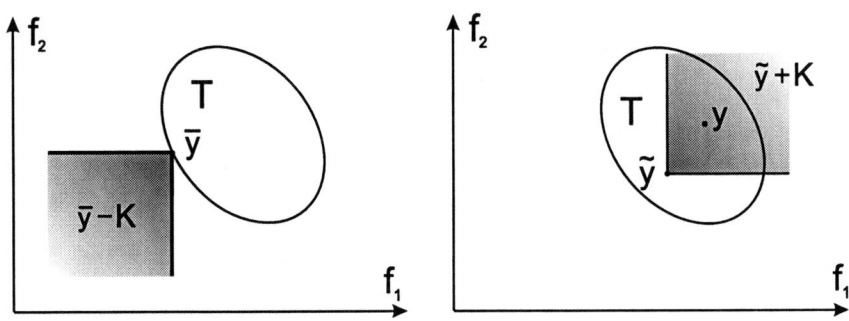

Fig. 1.1. K-minimal point $\bar{y}$. A point y dominated by the point $\tilde{y}$.

Using this concept we can define minimality for the multiobjective optimization problem (MOP).

Definition 1.3. A point $\bar{x} \in \Omega$ is a minimal solution (or non-dominated or efficient or K-minimal) of the multiobjective optimization problem (MOP) w.r.t. the ordering cone K if $f(\bar{x})$ is a K-minimal point of the set $f(\Omega)$. The set of all minimal solutions w.r.t. the cone K is denoted as $\mathcal{M}(f(\Omega), K)$. The image set of the set of minimal solutions

$$\mathcal{E}(f(\Omega), K) := \{f(x) \mid x \in \mathcal{M}(f(\Omega), K)\}$$

is called efficient set. A point $\bar{y} \in \mathcal{E}(f(\Omega), K)$ is called K-minimal, non-dominated, or efficient w.r.t. the cone K.

In Fig. 1.2 the example of a biobjective (bicriteria) optimization problem, i.e. $m = 2$, is given. The set Ω and $f(\Omega)$ as well as the pointed ordering cone is shown. The efficient set is drawn as thick line.

If there is a point $f(x) \in f(\Omega)$ with $f(x) - f(\bar{x}) \in K \setminus \{0_m\}$ then we say that $f(x)$ is dominated by $f(\bar{x})$ and x is dominated by $\bar{x}$ respectively.

For $K = \mathbb{R}^m_+$ the K-minimal points are also called Edgeworth-Pareto-minimal (EP-minimal) points according to Edgeworth, 1881

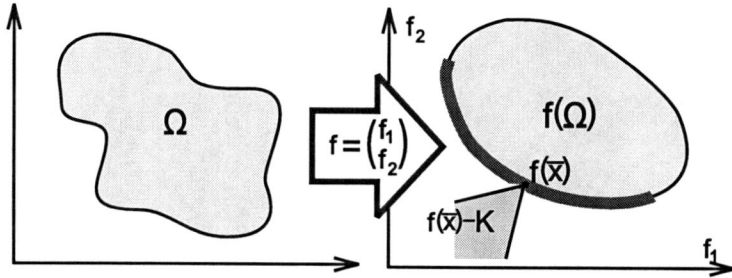

Fig. 1.2. K-minimality and efficient set for a biobjective optimization problem.

([59]), and Pareto, 1896 ([180]). For $K = \mathbb{R}^m_-$ we also speak of EP-maximal points.

In a partially ordered linear space there can exist points which are not comparable as e.g. the points $(1,2)$ and $(2,1)$ in $\mathbb{R}^2$ w.r.t. the natural ordering. That is the reason why in general (non-discrete) multiobjective optimization problems have an infinite number of solutions. The concern of this book is to determine the complete efficient set to make it available to the d. m.. Because an explicit calculation of all solutions is generally not possible we try to give a pointwise approximation of the efficient set with a high approximation quality.

In the case of a total ordering and if the multiobjective optimization problem is solvable there is only one minimal solution in the image space. An ordering is called total if for all $x, y \in \mathbb{R}^m$ either $x \leq y$ or $y \leq x$ is true. If the ordering is characterized by a convex cone $K \subset \mathbb{R}^m$ then it is total if and only if $K \cup (-K) = \mathbb{R}^m$ ([85, Theorem 2.1(4)]). For example the lexicographical ordering defined by the cone

$$K := \{y \in \mathbb{R}^m \mid \exists k \in \{1, \ldots, m\} \text{ with } y_i = 0 \text{ for } i < k \text{ and } y_k > 0\} \cup \{0_m\}$$

(see Fig. 1.3) is total. With respect to this ordering all points in $\mathbb{R}^m$ can be compared to each other and can be ordered. However a pointed convex cone cannot be closed and satisfy the total ordering property $K \cup (-K) = \mathbb{R}^m$ at the same time, see [85, p.4]. For example the ordering cone of the lexicographical ordering is not closed.

In this book we consider arbitrary closed pointed convex cones and thus the induced orderings $\leq_K$ are always not total but only partial. A special case of partial orderings is the natural ordering $\leq_m$, but we will not restrict ourselves to this special ordering and the related notion of EP-minimality. The EP-minimality notion is appropriate if the

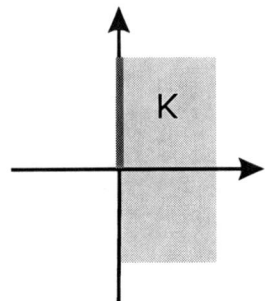

Fig. 1.3. Ordering cone of the lexicographical ordering for $m = 2$.

mathematical multiobjective optimization problem (MOP) adequately describes the real decision problem (w. r. t. the natural ordering). However there can exist preferences of the d. m. which cannot be mapped with the help of objective functions and constraints only, as it is demonstrated with the following example (see [230, Ex. 4.1]).

Example 1.4. We consider the multiobjective optimization problem

$$\min \begin{pmatrix} f_1(x) \\ f_2(x) \end{pmatrix} = \begin{pmatrix} x_1 \\ x_2 \end{pmatrix}$$

subject to the constraint

$$x \in \Omega \subset \mathbb{R}^2$$

with the constraint set Ω as given in Fig. 1.4.

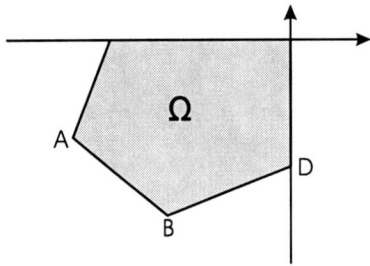

Fig. 1.4. Constraint set of Example 1.4.

We assume that the preferences of the d. m. are in such a way that there exists a third not explicitly known objective function $f_3(x) = x_2 - x_1$. Then a point $y^1 = f(x^1)$ is preferred to a point $y^2 = f(x^2)$ by the d. m. if

$$\Leftrightarrow \begin{pmatrix} f_1(x^1) \\ f_2(x^1) \\ f_3(x^1) \end{pmatrix} \leq_3 \begin{pmatrix} f_1(x^2) \\ f_2(x^2) \\ f_3(x^2) \end{pmatrix}$$

$$\Leftrightarrow \begin{pmatrix} y_1^1 \\ y_2^1 \\ y_2^1 - y_1^1 \end{pmatrix} \leq_3 \begin{pmatrix} y_1^2 \\ y_2^2 \\ y_2^2 - y_1^2 \end{pmatrix}$$

is satisfied. The ordering cone is thus defined by

$$K := \{y \in \mathbb{R}^2 \mid y_1 \geq 0, \ y_2 \geq 0, \ y_2 - y_1 \geq 0\} = \{y \in \mathbb{R}^2 \mid y_2 \geq y_1 \geq 0\}$$

(Fig. 1.5). We have $\mathcal{E}(f(\Omega), K) = \overline{AB} \cup \overline{BD}$ compared to $\mathcal{E}(f(\Omega), \mathbb{R}_+^2) = \overline{AB}$ (with $\overline{AB}$ and $\overline{BD}$ the line segment connecting the points A and B and B and D respectively).

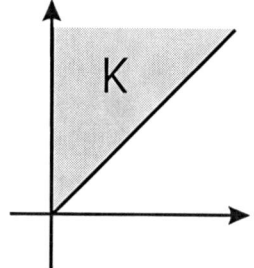

Fig. 1.5. Ordering cone of Example 1.4.

Many application problems demand the use of arbitrary closed pointed convex cones. For example in portfolio optimization in stock markets the domination orderings can be given by pointed convex polyhedral cones in $\mathbb{R}^m$ which are generated by more than m vectors ([4, p.692]). In Ex. 4.2 in [4] even a non-polyhedral cone, the so-called ice cream cone $K = \{x \in \mathbb{R}^3 \mid x_1 \geq \sqrt{x_2^2 + x_3^2}\}$ is mentioned for a domination structure (see also Example 1.22). In [116, 117, 185] the importance of polyhedral cones for modeling preferences of decision makers is discussed. Polyhedral cones are also used in [58]. We will present polyhedral and finitely generated cones including the definitions and a discussion in the following section.

Moreover, if the multiobjective optimization problem is only a tool for solving superordinate problems, ordering cones other than the natural ordering cone are of interest. In [85] a multiobjective optimization problem is constructed for solving scalar bilevel optimization problems. There, a non-convex cone is defined which can be expressed in a weakened form as the union of two convex cones, none of them equal to the positive orthant. Besides in part three of this book where the proposed methods are applied to solve multiobjective bilevel problems, an ordering different from the natural ordering appears, too. In [106, 134, 219] multiobjective optimization problems w.r.t. partial orderings defined by pointed convex cones are discussed for determining EP-minimal points of a multiobjective optimization problem. Because of all these considerations and examples we do not restrict ourselves to the natural ordering, like it is also done in [81, 96, 106, 136, 181, 230], and others.

A weaker minimality notion than K-minimality being of interest especially in theoretical considerations is the following:

Definition 1.5. Let K be a pointed ordering cone with $\text{int}(K) \neq \emptyset$. A point $\bar{x} \in \Omega$ is a weakly minimal solution of (MOP) w.r.t. K if

$$(\bar{y} - \text{int}(K)) \cap f(\Omega) = \emptyset.$$

The set of all weakly minimal solutions w.r.t. the cone K is denoted as $\mathcal{M}_w(f(\Omega), K)$. The image set of the set of weakly minimal points

$$\mathcal{E}_w(f(\Omega), K) := \{f(x) \mid x \in \mathcal{M}_w(f(\Omega), K)\}$$

is called the set of weakly efficient points w.r.t. the cone K.

We also speak of weakly EP-minimal points for $K = \mathbb{R}^m_+$.

The weakly K-minimal points are the minimal points w.r.t. the cone $\text{int}(K) \cup \{0_m\}$. Therefore, considering the following lemma, we have in the case that $\text{int}(K)$ is nonempty that

$$\mathcal{M}(f(\Omega), K) \subset \mathcal{M}_w(f(\Omega), K)$$

and

$$\mathcal{E}(f(\Omega), K) \subset \mathcal{E}_w(f(\Omega), K).$$

Lemma 1.6. ([190, Prop. 3.1.1]) Let K_1 and K_2 be nonempty convex cones with $K_1 \subset K_2$. Then we have for the set of minimal solutions

$$\mathcal{M}(f(\Omega), K_2) \subset \mathcal{M}(f(\Omega), K_1).$$

A similar result can be stated for the set of weakly minimal points.

1.1 Basic Concepts 11

Lemma 1.7. *Let K_1 and K_2 be pointed ordering cones with nonempty interior and with $K_1 \subset K_2$. Then we have for the set of weakly minimal solutions*
$$\mathcal{M}_w(f(\Omega), K_2) \subset \mathcal{M}_w(f(\Omega), K_1).$$

Proof. As $K_1 \subset K_2$ it follows $\text{int}(K_1) \subset \text{int}(K_2)$ and therefore
$$\begin{aligned}\mathcal{M}_w(f(\Omega), K_2) &= \mathcal{M}(f(\Omega), \text{int}(K_2) \cup \{0_m\}) \\ &\subset \mathcal{M}(f(\Omega), \text{int}(K_1) \cup \{0_m\}) \\ &= \mathcal{M}_w(f(\Omega), K_1).\end{aligned}$$
□

We now give a short excursus how the definition of weakly minimal points can be extended for the case of an empty interior of the ordering cone K. We give the definition for a general linear space X but for simplicity the reader can replace X by $\mathbb{R}^m$. First we introduce the algebraic interior of a nonempty set $K \subset X$, also called core of K ([124, p.6]),
$$\text{cor}(K) := \{\bar{x} \in K \mid \text{ for every } x \in X \text{ there exists a } \bar{\lambda} > 0 \\ \text{with } \bar{x} + \lambda x \text{ for all } \lambda \in [0, \bar{\lambda}]\}.$$

If the set K is a convex cone with $\text{cor}(K) \neq \emptyset$, then $\text{cor}(K) \cup \{0_X\}$ is a convex cone, too. For K a convex cone in a topological space (as $X = \mathbb{R}^m$) and $\text{int}(K) \neq \emptyset$ it is $\text{int}(K) = \text{cor}(K)$.

We are interested in the relative algebraic interior also called intrinsic core. Let $L(K)$ denote the smallest subspace of X containing the set $K \subset X$ and let K be a convex cone. Then the intrinsic core is defined as ([181, p.501], [2, p.517])
$$\text{icr}(K) := \{\bar{x} \in K \mid \text{ for every } x \in L(K) \text{ there exists a } \bar{\lambda} > 0 \\ \text{with } \bar{x} + \lambda x \text{ for all } \lambda \in [0, \bar{\lambda}]\}.$$

If K is a convex cone with $\text{icr}(K) \neq \emptyset$, then $\text{icr}(K) \cup \{0_X\}$ is a convex cone, too. Of course it is $\text{cor}(K) \subset \text{icr}(K) \subset K$. If $K \subset \mathbb{R}^m$ is a convex cone with $\text{icr}(K) \neq \emptyset$ it is according to Lemma 1.6 of course
$$\mathcal{M}(f(\Omega), K) \subset \mathcal{M}(f(\Omega), \text{icr}(K) \cup \{0_m\}). \tag{1.2}$$

If further $\text{int}(K) \neq \emptyset$, then due to $\text{int}(K) \subset \text{icr}(K)$ we conclude

$$\mathcal{M}(f(\Omega), \mathrm{icr}(K) \cup \{0_m\}) \subset \mathcal{M}(f(\Omega), \mathrm{int}(K) \cup \{0_m\}) = \mathcal{M}_w(f(\Omega), K). \tag{1.3}$$

Weak minimality can be defined w. r. t. the algebraic interior or w. r. t. the relative algebraic interior, too, see e. g. [120], [124, p.109] or [181]. This is especially interesting if the ordering cone has an empty (topological) interior.

Example 1.8. We consider the ordering cone

$$K = \{y \in \mathbb{R}^3 \mid y_1 \geq 0, \; y_2 \geq 0, \; y_3 = 0\} \subset \mathbb{R}^3.$$

It is $K = \mathbb{R}_+^2 \times \{0\}$. Then $\mathrm{int}(K) = \emptyset$ and we cannot define weak minimality. However it is $\mathrm{icr}(K) = \{y \in \mathbb{R}^3 \mid y_1 > 0, \; y_2 > 0, \; y_3 = 0\}$ and we can define a modified weak minimality, i. e. minimality w. r. t. $\mathrm{icr}(K) \cup \{0_3\}$.

This concept based on the intrinsic core is used in the paper of Pascoletti and Serafini ([181]) on which we base our scalarization approach in the following section. Besides we need this modified definition of weak minimality in the last part of this book where we discuss multiobjective bilevel optimization problems. There, an ordering cone appears with an empty interior.

Various other and stronger minimality notions are defined in the literature (see for instance [124] and the references therein) as e. g. proper minimality (different definitions can be found in [12, 21, 90, 107, 144, 121], and others). We mention here only the definition of proper minimality according to Geoffrion ([90, 190]) for the ordering cone $K = \mathbb{R}_+^m$.

Definition 1.9. Let $K = \mathbb{R}_+^m$. A point $\bar{x}$ is a properly efficient solution of the multiobjective optimization problem (MOP) if it is EP-minimal and if there is some real $M > 0$ such that for each $i \in \{1, \ldots, m\}$ and each $x \in \Omega$ satisfying $f_i(x) < f_i(\bar{x})$, there exists at least one $j \in \{1, \ldots, m\}$ such that $f_j(\bar{x}) < f_j(x)$ and

$$\frac{f_i(\bar{x}) - f_i(x)}{f_j(x) - f_j(\bar{x})} \leq M.$$

The relation between properly efficient solutions and stable solutions of scalarizations w. r. t. perturbations of the constraint set is considered in [169].

Next we present some useful calculation rules (see [190, pp.34f]).

1.1 Basic Concepts 13

Lemma 1.10. *Let a nonempty ordering cone K, a scalar $\alpha > 0$, and the sets $f(\Omega), \tilde{f}(\tilde{\Omega}) \subset \mathbb{R}^m$, be given. Then it is:*

a) $\mathcal{E}(\alpha f(\Omega), K) = \alpha \mathcal{E}(f(\Omega), K)$,
b) $\mathcal{E}(f(\Omega) + \tilde{f}(\tilde{\Omega}), K) \subset \mathcal{E}(f(\Omega), K) + \mathcal{E}(\tilde{f}(\tilde{\Omega}), K)$,
c) $\mathcal{E}(f(\Omega), K) \supset \mathcal{E}(f(\Omega) + K, K)$.
d) If additionally the cone K is pointed then

$$\mathcal{E}(f(\Omega), K) = \mathcal{E}(f(\Omega) + K, K).$$

The following rule will be of interest in part three of this book dealing with bilevel optimization.

Lemma 1.11. *For two sets $A^0, A^1 \subset \mathbb{R}^n$, and a vector-valued function $f: \mathbb{R}^n \to \mathbb{R}^m$, consider the sets*

$$A := A^0 \cup A^1 \quad \text{and}$$
$$\tilde{A} := \mathcal{M}(f(A^0), \mathbb{R}^m_+) \cup A^1.$$

Let $f(A^0)$ be compact. Then it is $\mathcal{M}(f(A), \mathbb{R}^m_+) = \mathcal{M}(f(\tilde{A}), \mathbb{R}^m_+)$.

Proof. First we show $\mathcal{M}(f(A), \mathbb{R}^m_+) \subset \mathcal{M}(f(\tilde{A}), \mathbb{R}^m_+)$. For that we assume $\bar{x} \in \mathcal{M}(f(A), \mathbb{R}^m_+)$. Then there exists no point $x' \in A$ with

$$f(x') \leq f(\bar{x}) \text{ and } f(x') \neq f(\bar{x}). \tag{1.4}$$

We have $\bar{x} \in A = A^0 \cup A^1$. If $\bar{x} \in A^0$ then there is no point $x' \in A^0 \subset A$ with (1.4) and hence $\bar{x} \in \mathcal{M}(f(A^0), \mathbb{R}^m_+) \subset \tilde{A}$. For $\bar{x} \in A^1$ we have $\bar{x} \in \tilde{A}$, too. Because of $\tilde{A} \subset A$ there also exists no $x' \in \tilde{A}$ with (1.4) and hence $\bar{x} \in \mathcal{M}(f(\tilde{A}), \mathbb{R}^m_+)$.

It remains to show $\mathcal{M}(f(\tilde{A}), \mathbb{R}^m_+) \subset \mathcal{M}(f(A), \mathbb{R}^m_+)$. For that we assume $\bar{x} \in \mathcal{M}(f(\tilde{A}), \mathbb{R}^m_+)$, i.e. there is no $x' \in \tilde{A}$ with (1.4). We have $\bar{x} \in \mathcal{M}(f(A^0), \mathbb{R}^m_+) \cup A^1$. For $\bar{x} \in \mathcal{M}(f(A^0), \mathbb{R}^m_+)$ there exists no $x' \in A^0$ with (1.4), too. Because of $A = \tilde{A} \cup A^0$ we conclude $\bar{x} \in \mathcal{M}(f(A), \mathbb{R}^m_+)$. And for $\bar{x} \in A^1$ we assume $\bar{x} \notin \mathcal{M}(f(A), \mathbb{R}^m_+)$. Then there is a $x' \in A \setminus \tilde{A} = A^0 \setminus \mathcal{M}(f(A^0), \mathbb{R}^m_+)$ with (1.4). As the set $f(A^0)$ is compact it holds according to [190, Theorem 3.2.10] $f(A^0) \subset \mathcal{E}(f(A^0), \mathbb{R}^m_+) + \mathbb{R}^m_+$. Because of $x' \in A^0$ and $x' \notin \mathcal{M}(f(A^0), \mathbb{R}^m_+)$ there exists a $x^0 \in \mathcal{M}(f(A^0), \mathbb{R}^m_+)$ with $f(x^0) \leq f(x')$ resulting in $f(x^0) \leq f(\bar{x})$ and $f(x^0) \neq f(\bar{x})$. Due to $x^0 \in \tilde{A}$ this is a contradiction to $\bar{x} \in \mathcal{M}(f(\tilde{A}), \mathbb{R}^m_+)$. Hence $\bar{x} \in \mathcal{M}(f(A), \mathbb{R}^m_+)$. □

The following theorem (see [94, Theorem 2.9]) shows that it is sufficient to consider only the boundary $\partial f(\Omega)$ of the set $f(\Omega)$ for determining all efficient points:

Theorem 1.12. *Let K be a nonempty ordering cone with $K \neq \{0_m\}$. Then*
$$\mathcal{E}(f(\Omega), K) \subset \partial f(\Omega).$$

This is true for the weakly efficient points, too.

Theorem 1.13. *Let K be a pointed ordering cone with $\text{int}(K) \neq \emptyset$. Then*
$$\mathcal{E}_w(f(\Omega), K) \subset \partial f(\Omega).$$

Proof. For an $\bar{y} \in \mathcal{E}_w(f(\Omega), K)$ we assume $\bar{y} \in f(\Omega) \setminus \partial f(\Omega) = \text{int}(f(\Omega))$. Then there exists a $\delta > 0$ and an open ball $B = \{y \in \mathbb{R}^m \mid \|y\| < \delta\}$ with $\bar{y} + B \subset f(\Omega)$. Let $k \in \text{int}(K)$. Then there is a $\lambda < 0$ with $\lambda k \in B$ and it is $\bar{y} + \lambda k \in f(\Omega)$. Because K is a cone it is $\lambda k \in -\text{int}(K)$ and hence we have
$$\bar{y} + \lambda k \in f(\Omega) \cap (\bar{y} - \text{int}(K))$$
in contradiction to $\bar{y}$ weakly K-minimal. □

As it is known from scalar optimization there is also the notion of local minimality.

Definition 1.14. *Let K be a closed pointed ordering cone with $\text{int}(K) \neq \emptyset$.*

A point $\bar{x} \in \Omega$ is a *local minimal solution* of the multiobjective optimization problem (MOP) w. r. t. the ordering cone K if there is a neighborhood U of $\bar{x}$ such that there is no $y \in f(\Omega \cap U) \setminus \{f(\bar{x})\}$ with $f(\bar{x}) \in y + K$.

A point $\bar{x} \in \Omega$ is a *locally weakly minimal solution* of the multiobjective optimization problem (MOP) w. r. t. the ordering cone K if there is a neighborhood U of $\bar{x}$ such that there is no $y \in f(\Omega \cap U)$ with $f(\bar{x}) \in y + \text{int}(K)$.

In applications and in numerical procedures the notion of ε-EP-minimality (see for instance [125, 154, 212]) is useful:

Definition 1.15. *Let $\varepsilon \in \mathbb{R}^m$ with $\varepsilon_i > 0$, $i = 1, \ldots, m$, be given. A point $\bar{x} \in \Omega$ is an ε-EP-minimal solution of the multiobjective optimization problem (MOP) if there is no $x \in \Omega$ with*

$$f_i(x) + \varepsilon_i \le f_i(\bar{x}) \quad \text{for all } i \in \{1, \ldots, m\}$$

and

$$f_j(x) + \varepsilon_j < f_j(\bar{x}) \quad \text{for at least one } j \in \{1, \ldots, m\}.$$

In Fig. 1.6 the image set of the ε-EP-minimal solutions of a biobjective optimization problem is shown (compare [125]).

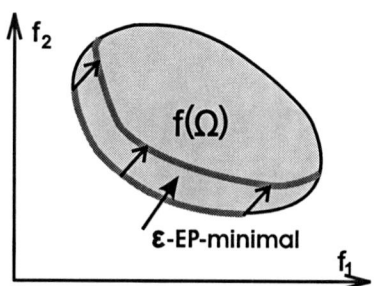

Fig. 1.6. ε-EP-minimality for a biobjective optimization problem.

For ε-minimal solutions w.r.t. arbitrary ordering cones K see for instance [73].

1.2 Polyhedral Ordering Cones

Special cones are the so-called polyhedral cones. Their properties can be used for simplifying the solving of a multiobjective optimization problem.

Definition 1.16. ([190, Def. 2.1.7, 2.1.8])

a) A set $K \subset \mathbb{R}^m$ is a convex polyhedral cone if K can be represented by
$$K = \{x \in \mathbb{R}^m \mid (\bar{k}^i)^\top x \ge 0, \quad i = 1, \ldots, s\}$$
with $s \in \mathbb{N}$ and vectors $\bar{k}^i \in \mathbb{R}^m$, $i = 1, \ldots, s$.

b) A set $K \subset \mathbb{R}^m$ is a finitely generated convex cone if there are vectors $a^1, a^2, \ldots, a^s$, $s \in \mathbb{N}$, in $\mathbb{R}^m$ such that K can be described by
$$K = \{x \in \mathbb{R}^m \mid x = \sum_{i=1}^{s} \alpha_i a^i, \ \alpha_i \ge 0, \ i = 1, \ldots, s\}.$$

Lemma 1.17. *([190, Prop. 2.1.12], [219, Lemma 2.1, 2.2]) A convex cone K is polyhedral if and only if it is finitely generated.*

If the cone K can be represented by

$$K = \{x \in \mathbb{R}^m \mid (\overline{k}^i)^\top x \geq 0, \quad i = 1, \ldots, s\}$$

($s \in \mathbb{N}$), then we say that the cone K is generated or induced by the matrix

$$\overline{K} := \begin{pmatrix} (\overline{k}^1)^\top \\ \vdots \\ (\overline{k}^s)^\top \end{pmatrix}$$

and we have $K = \{x \in \mathbb{R}^m \mid \overline{K}x \geq_s 0_s\}$.

If the cone K is generated by the matrix $\overline{K}$ and if $\text{kernel}(\overline{K}) = \{0_m\}$, then K is pointed and the related partial ordering is antisymmetric. For example the pointed cone defining the natural ordering is polyhedral and is induced by the m-dimensional identity-matrix E_m.

The task of finding K-minimal points of a multiobjective optimization problem w.r.t. a pointed polyhedral ordering cone generated by the matrix $\overline{K} \in \mathbb{R}^{s \times m}$, can be reduced to the determination of EP-minimal points of the multiobjective optimization problem

$$\min \overline{K} f(x)$$

subject to the constraint

$$x \in \Omega$$

with s objective functions $(\overline{k}^1)^\top f(x), (\overline{k}^2)^\top f(x), \ldots, (\overline{k}^s)^\top f(x)$, see [190, Lemma 2.3.4]:

Lemma 1.18. *We consider the multiobjective optimization problem (MOP) with a polyhedral ordering cone K represented by*

$$K = \{x \in \mathbb{R}^m \mid \overline{K}x \geq_s 0_s\}$$

with $\overline{K} \in \mathbb{R}^{s \times m}$ and $\text{kernel}(\overline{K}) = \{0_m\}$. Then

$$\mathcal{E}(f(\Omega), K) = \{y \in f(\Omega) \mid \overline{K}y \in \mathcal{E}(\overline{K}f(\Omega), \mathbb{R}_+^s)\}$$

and

$$\mathcal{M}(f(\Omega), K) = \mathcal{M}(\overline{K}f(\Omega), \mathbb{R}_+^s)$$

with $\overline{K}f(\Omega) := \{\overline{K}y \in \mathbb{R}^s \mid y \in f(\Omega)\}$.

1.2 Polyhedral Ordering Cones

Hence, if the ordering cone is polyhedral and induced by a $s \times m$-matrix, we can reduce the problem of finding K-minimal points of a multiobjective optimization problem with m objective functions to the problem of finding EP-minimal points of a multiobjective optimization problem with s criteria. However if $s > m$ the new problem is getting more complex.

Example 1.19. We consider the determination of K-minimal points of the optimization problem

$$\min_{x \in \Omega} \begin{pmatrix} f_1(x) \\ f_2(x) \\ f_3(x) \end{pmatrix} \qquad (1.5)$$

with the ordering cone

$$K = \{y \in \mathbb{R}^3 \mid \begin{pmatrix} 1 & 0 & 1 \\ -1 & 0 & 1 \\ 0 & -1 & 1 \\ 0 & 1 & 1 \end{pmatrix} y \geq_4 0_4\}.$$

It is $K = \{y \in \mathbb{R}^3 \mid -y_3 \leq y_1 \leq y_3,\ -y_3 \leq y_2 \leq y_3,\ y_3 \geq 0\}$ a pyramid with apex in the origin.

According to Lemma 1.18 a point $\bar{x} \in \Omega$ is a K-minimal solution of (1.5) if and only if $\bar{x}$ is EP-minimal for

$$\min_{x \in \Omega} \begin{pmatrix} f_1(x) + f_3(x) \\ -f_1(x) + f_3(x) \\ -f_2(x) + f_3(x) \\ f_2(x) + f_3(x) \end{pmatrix}.$$

We will discuss this special property of multiobjective optimization problems with finitely generated ordering cones again in Chap. 3 in the context of the sensitivity studies and in Chap. 5. In the bicriteria case ($m = 2$) every ordering cone is finitely generated.

Lemma 1.20. Let $K \subset \mathbb{R}^2$ be a closed pointed ordering cone with $K \neq \{0_2\}$. Then K is polyhedral and there is either a $k \in \mathbb{R}^2 \setminus \{0_2\}$ with

$$K = \{\lambda k \mid \lambda \geq 0\}$$

or there are $l^1, l^2 \in \mathbb{R}^2 \setminus \{0_2\}$, l^1, l^2 linearly independent, and $\tilde{l}^1, \tilde{l}^2 \in \mathbb{R}^2 \setminus \{0_2\}$, $\tilde{l}^1, \tilde{l}^2$ linearly independent, with

$$K = \{y \in \mathbb{R}^2 \mid l^{1\top} y \geq 0, \; l^{2\top} y \geq 0\}$$
$$= \{y \in \mathbb{R}^2 \mid y = \lambda^1 \tilde{l}^1 + \lambda^2 \tilde{l}^2, \; \lambda^1, \lambda^2 \geq 0\}.$$

Proof. We give a constructive proof with which we can determine k and l^1, l^2 and $\tilde{l}^1, \tilde{l}^2$ respectively by solving simple optimization problems. We start by solving

$$\min \varphi$$

subject to the constraints

$$\begin{pmatrix} \cos \varphi \\ \sin \varphi \end{pmatrix} \in K, \qquad (1.6)$$
$$\varphi \in [0, 2\pi]$$

with minimal solution φ^1. Next, assuming $\varphi^1 \neq 0$, we solve

$$\max \varphi$$

subject to the constraints

$$\begin{pmatrix} \cos \varphi \\ \sin \varphi \end{pmatrix} \in K, \qquad (1.7)$$
$$\varphi \in [0, 2\pi]$$

with maximal solution φ^2. Because the cone K is closed and non-trivial there always exists a solution of (1.6) and (1.7) and as the cone K is also pointed we have $\varphi^2 \in [\varphi^1, \varphi^1 + \pi[$.
For $\varphi^1 = \varphi^2$ this results in $K = \{\lambda k \mid k = (\cos \varphi^1, \sin \varphi^1)^\top, \; \lambda \geq 0\}$.
For $\varphi^1 \neq \varphi^2$ we get due to the convexity of K

$$K = \{y \in \mathbb{R}^2 \mid y = \lambda (\cos \varphi, \sin \varphi)^\top, \; \lambda \geq 0, \; \varphi \in [\varphi^1, \varphi^2]\}$$
$$= \{y \in \mathbb{R}^2 \mid y = \lambda^1 (\cos \varphi^1, \sin \varphi^1)^\top + \lambda^2 (\cos \varphi^2, \sin \varphi^2)^\top, \lambda^1, \lambda^2 \geq 0\}.$$

In the same manner we can handle the case $\varphi^1 = 0$ by solving the problems (1.6) and (1.7) w.r.t. $\varphi \in [\pi, 3\pi]$ instead of $\varphi \in [0, 2\pi]$.
We have

$$\tilde{l}^1 := (\cos \varphi^1, \sin \varphi^1)^\top \text{ and } \tilde{l}^2 := (\cos \varphi^2, \sin \varphi^2)^\top,$$

with $\tilde{l}^1$, $\tilde{l}^2$ linearly independent because of the pointedness of K. With appropriate orthogonal vectors l^1, l^2 to $\tilde{l}^1$, $\tilde{l}^2$ we obtain

$$K = \{y \in \mathbb{R}^2 \mid l^{1\top} y \geq 0, \ l^{2\top} y \geq 0\}.$$

□

For instance the cone $\mathbb{R}^2_+$ is finitely generated by $l^1 = (1,0)^\top$ and $l^2 = (0,1)^\top$. In general the statement of the theorem is not true in $\mathbb{R}^m$, $m \geq 3$. This is shown by the following two examples, the ordering cone of the Löwner partial ordering and the so-called ice cream cone. These cones also demonstrate, that there are important cones which are non-polyhedral.

Example 1.21. The Löwner partial ordering ([155]) in the space $\mathcal{S}^n$ of symmetric $n \times n$-matrices is defined by the convex cone $\mathcal{S}^n_+$ of positive semidefinite $n \times n$-matrices

$$\mathcal{S}^n_+ := \{A \in \mathcal{S}^n \mid x^\top A x \geq 0 \quad \text{for all } x \in \mathbb{R}^n\}.$$

With the help of a vector-valued map the space $\mathcal{S}^n$ can be mapped in the space $\mathbb{R}^{n(n+1)/2}$ (see e.g. in [218, Def. 2.3] the map named svec($\cdot$)). For the case $n = 2$ this map is the following:

$$A = \begin{pmatrix} x & y \\ y & z \end{pmatrix} \in \mathcal{S}^2 \quad \mapsto \quad a = \begin{pmatrix} x \\ y \\ z \end{pmatrix} \in \mathbb{R}^3.$$

Based on this map the Löwner ordering cone can be described in the space $\mathbb{R}^3$ by the cone

$$\begin{aligned} K &= \{(x,y,z) \in \mathbb{R}^3 \mid x + z \geq 0, \ xz \geq y^2\} \\ &= \{(x,y,z) \in \mathbb{R}^3 \mid x \geq 0, \ z \geq 0, \ xz \geq y^2\}. \end{aligned}$$

Thus $A \in \mathcal{S}^2_+$ if and only if $a \in K$. The closed pointed convex cone K (Fig. 1.7) is not finitely generated and hence not polyhedral.

Example 1.22. In [4, Ex. 4.2] (with reference to [5]) a problem in portfolio optimization is discussed, for which the domination structure in the three dimensional portfolio space is defined by the so-called ice cream cone

$$K := \{(x,y,z) \in \mathbb{R}^3 \mid x \geq \sqrt{y^2 + z^2}\},$$

also called second-order cone. This is again a closed pointed convex cone which is not finitely generated and thus non-polyhedral.

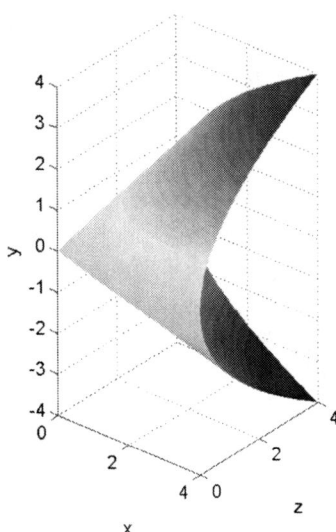

Fig. 1.7. Löwner ordering cone of $\mathcal{S}^2$ mapped to $\mathbb{R}^3$.

For general non-polyhedral cones we cannot give an equation like the one in Lemma 1.18 as it is shown in [72, Theorem 2].

2

Scalarization Approaches

For determining solutions of the multiobjective optimization problem (MOP)

$$\text{(MOP)} \quad \min f(x)$$
$$\text{subject to the constraints}$$
$$g(x) \in C,$$
$$h(x) = 0_q,$$
$$x \in S$$

with the constraint set $\Omega = \{x \in S \mid g(x) \in C,\ h(x) = 0_q\}$ a widespread approach is the transformation of this problem to a scalar-valued parameter dependent optimization problem. This is done for instance in the weighted sum method ([245]). There the scalar problems

$$\min_{x \in \Omega} \sum_{i=1}^{m} w_i f_i(x)$$

with weights $w \in K^* \setminus \{0_m\}$ and K^* the dual cone to the cone K, i.e. $K^* = \{y^* \in \mathbb{R}^m \mid (y^*)^\top y \geq 0 \text{ for all } y \in K\}$, are solved. Another scalarization especially for calculating EP-minimal points is based on the minimization of only one of the m objectives while all the other objectives are transformed into constraints by introducing upper bounds. This scalarization is called ε-constraint method ([98, 159]) and is given by

$$\min f_k(x)$$
$$\text{subject to the constraints}$$
$$f_i(x) \leq \varepsilon_i, \quad i \in \{1, \ldots, m\} \setminus \{k\},$$
$$x \in \Omega.$$
(2.1)

Here the parameters are the upper bounds $\varepsilon_i, i \in \{1,\ldots,m\} \setminus \{k\}$ for a $k \in \{1,\ldots,m\}$. Surveys about different scalarization approaches can be found in [60, 112, 124, 138, 165, 189]. Other solution approaches use e. g. stochastic methods as it is done by Schäffler, Schultz and Weinzierl ([194]) or evolutionary algorithms (surveys for these types of methods can be found in [112, p.19] and in [41, 42, 31, 228, 246]). In this book only procedures based on a scalarization of the multiobjective optimization problem are considered.

By solving the scalar problems for a variety of parameters for instance for different weights, several solutions of the multiobjective optimization problem are generated. In the last decades the main focus was on finding one minimal solution e. g. by interactive methods ([166, 165]) whereas objective numerical calculations alternate with subjective decisions done by the decision maker. Based on much better computer performances it is now possible to represent the whole efficient set. Having the whole solution set available the decision maker gets a useful insight in the problem structure. For engineering tasks, it is especially interesting to have all design alternatives available ([119]). The aim is to generate an approximation of the whole efficient set as it is the aim for instance in [40, 47, 48, 81, 106, 110, 111, 164, 196] and many more.

The information provided by this approximation depends mainly on the quality of the approximation. Many approximation points cause a high numerical effort, however approximations with only few points neglect large areas of the efficient set. Thus, the aim of this book is to generate an approximation with a high quality.

A wide variety of scalarizations exist based on which one can determine single approximation points. However not all methods are appropriate for non-convexity or arbitrary partial orderings. For instance the weighted sum method has the disadvantage that it is in general only possible for convex problems to determine all efficient points by an appropriate parameter choice (see [39, 138]). The ε-constraint method as given in (2.1) is only suited for the calculation of EP-minimal points. Yet problems arising in applications are often non-convex. Further it is also of interest to consider more general partial orderings than the natural ordering. Thus we concentrate on a scalarization by Pascoletti and Serafini, 1984 ([181]) which we present and discuss in the following sections. An advantage of this scalarization is that many other scalarization approaches as the mentioned weighted sum method or the ε-constraint method are included in this more general formulation. The relationship to other scalarization problems are examined in the last section of this chapter.

2.1 Pascoletti-Serafini Scalarization

Pascoletti and Serafini propose the following scalar optimization problem with parameters $a \in \mathbb{R}^m$ and $r \in \mathbb{R}^m$ for determining minimal solutions of (MOP) w. r. t. the cone K:

$$\text{(SP(a,r))} \quad \min t$$

$$\text{subject to the constraints}$$
$$a + tr - f(x) \in K,$$
$$g(x) \in C,$$
$$h(x) = 0_q,$$
$$t \in \mathbb{R}, \ x \in S.$$

This problem has the parameter dependent constraint set

$$\Sigma(a, r) := \{(t, x) \in \mathbb{R}^{n+1} \mid a + tr - f(x) \in K, \ x \in \Omega\}.$$

We assume that the cone K is a nonempty closed pointed convex cone. The formulation of this scalar optimization problem corresponds to the definition of K-minimality. A point $\bar{x} \in \Omega$ with $\bar{y} = f(\bar{x})$ is K-minimal if

$$(\bar{y} - K) \cap f(\Omega) = \{\bar{y}\},$$

(see Fig. 2.1 for $m = 2$ and $K = \mathbb{R}^2_+$). If we rewrite the problem (SP(a, r)) as follows

$$\min t$$
$$\text{subject to the constraints}$$
$$f(x) \in a + tr - K,$$
$$x \in \Omega,$$
$$t \in \mathbb{R},$$

we see that for solving this problem the ordering cone $-K$ is moved in direction $-r$ on the line $a + tr$ starting in the point a till the set $(a + tr - K) \cap f(\Omega)$ is reduced to the empty set. The smallest value $\bar{t}$ for which $(a + \bar{t}r - K) \cap f(\Omega) \neq \emptyset$ is the minimal value of (SP(a, r)) (see Fig. 2.2 with $m = 2$ and $K = \mathbb{R}^2_+$).

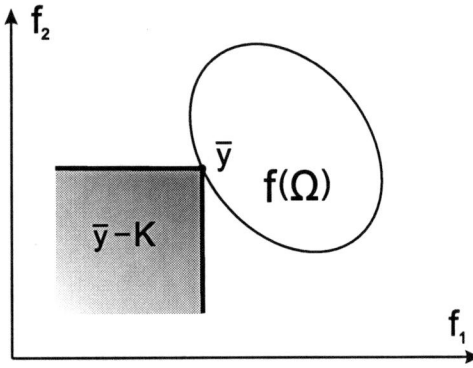

Fig. 2.1. K-minimality.

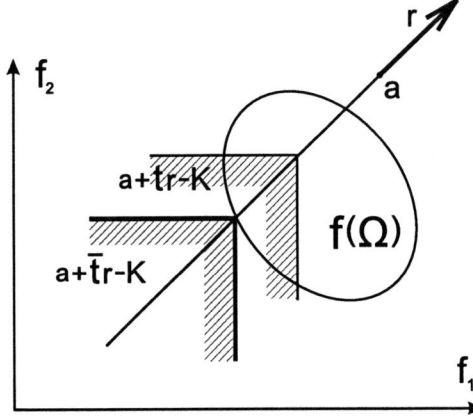

Fig. 2.2. Moving the ordering cone in the Pascoletti-Serafini problem.

The scalar problem (SP(a,r)) features all important properties a scalarization approach for determining minimal solutions of (MOP) should have. If $(\bar{t}, \bar{x})$ is a minimal solution of (SP(a,r)) then the point $\bar{x}$ is an at least weakly K-minimal solution of the multiobjective optimization problem (MOP) and by a variation of the parameters $(a, r) \in \mathbb{R}^m \times \mathbb{R}^m$ all K-minimal points of (MOP) can be found as solutions of (SP(a,r)). We will discuss these important properties among others in the following section.

Problem (SP(a,r)) is also discussed by Helbig in [104]. He interprets the point a as a reference point and the parameter r as a direction. For $r \in \text{int}(\mathbb{R}^m_+)$ this corresponds to the interpretation of r as a weighting of the objective functions with the weights $w_i := \frac{1}{r_i}$, $i = 1, \ldots, m$

(compare with the weighted Chebyshev norm). Then for a minimal solution $\bar{x}$ of the scalar problem the point $f(\bar{x}) = \bar{y}$ is the (weakly) K-minimal point (see Theorem 2.1,c)) which is closest to the reference point. The Pascoletti-Serafini problem is also related to a scalarization introduced by Gerstewitz in [91] as well as to the problem discussed in [92, 237] by Tammer, Weidner and Winkler. Further, in [74] Engau and Wiecek examine the Pascoletti-Serafini scalarization concerning ε-efficiency.

Pascoletti and Serafini allow for the parameter r only $r \in L(K)$ with $L(K)$ the smallest linear subspace in $\mathbb{R}^m$ including K. Sterna-Karwat discusses in [213] and [214] also this problem. Helbig ([104, 106]) assumes $r \in \text{rint}(K)$, i.e. he assumes r to be an element of the relative interior of the closed pointed convex cone K.

In [106] Helbig varies not only the parameters a and r, but he also varies the cone K. For $s \in K^*$ he defines parameter dependent cones $K(s)$ with $K \subset K(s)$. For these cones he solves the scalar problems (SP(a,r)). The solutions are then weakly $K(s)$-minimal, yet w.r.t. the cone K they are even minimal. However in this book we concentrate on a variation of the parameters a and r. We will see that by an appropriate controlling of the parameters high-quality approximations in the sense of nearly equidistant approximations of the efficient set can be generated.

2.2 Properties of the Pascoletti-Serafini Scalarization

We examine the Pascoletti-Serafini problem in this section more detailed and we start with the main properties of this scalarization (see also [181]). We assume again that K is a nonempty closed pointed ordering cone in $\mathbb{R}^m$.

Theorem 2.1. *We consider the scalar optimization problem (SP(a,r)) to the multiobjective optimization problem (MOP). Let $\text{int}(K) \neq \emptyset$.*

a) *Let $\bar{x}$ be a weakly K-minimal solution of the multiobjective optimization problem (MOP), then $(0, \bar{x})$ is a minimal solution of (SP(a,r)) for the parameter $a := f(\bar{x})$ and for arbitrary $r \in \text{int}(K)$.*

b) *Let $\bar{x}$ be a K-minimal solution of the multiobjective optimization problem (MOP), then $(0, \bar{x})$ is a minimal solution of (SP(a,r)) for the parameter $a := f(\bar{x})$ and for arbitrary $r \in K \setminus \{0_m\}$.*

26   2 Scalarization Approaches

c) Let $(\bar{t}, \bar{x})$ be a minimal solution of the scalar problem (SP(a, r)), then $\bar{x}$ is a weakly K-minimal solution of the multiobjective optimization problem (MOP) and $a + \bar{t} r - f(\bar{x}) \in \partial K$ with ∂K the boundary of the cone K.

d) Let $\bar{x}$ be a locally weakly K-minimal solution of the multiobjective optimization problem (MOP), then $(0, \bar{x})$ is a local minimal solution of (SP(a, r)) for the parameter $a := f(\bar{x})$ and for arbitrary $r \in \text{int}(K)$.

e) Let $\bar{x}$ be a locally K-minimal solution of the multiobjective optimization problem (MOP), then $(0, \bar{x})$ is a local minimal solution of (SP(a, r)) for the parameter $a := f(\bar{x})$ and for arbitrary $r \in K \setminus \{0_m\}$.

f) Let $(\bar{t}, \bar{x})$ be a local minimal solution of (SP(a, r)), then $\bar{x}$ is a locally weakly K-minimal solution of the multiobjective optimization problem (MOP) and $a + \bar{t} r - f(\bar{x}) \in \partial K$.

Proof. a) Set $a = f(\bar{x})$ and choose $r \in \text{int}(K)$ arbitrarily. Then the point $(0, \bar{x})$ is feasible for (SP(a, r)). It is also minimal, because otherwise there exists a feasible point (t', x') with $t' < 0$ and a $k' \in K$ with
$$a + t' r - f(x') = k'.$$
Hence we have $f(\bar{x}) = f(x') + k' - t' r$. It is $k' - t' r \in \text{int}(K)$ and thus it follows $f(\bar{x}) \in f(x') + \text{int}(K)$ in contradiction to $\bar{x}$ weakly K-minimal.

b) Set $a = f(\bar{x})$ and choose $r \in K \setminus \{0_m\}$ arbitrarily. Then the point $(0, \bar{x})$ is feasible for (SP(a, r)). It is also a minimal solution because otherwise there exists a scalar $t' < 0$ and a point $x' \in \Omega$, with (t', x') feasible for (SP(a, r)), and a $k' \in K$ with $a + t' r - f(x') = k'$. This leads to
$$f(\bar{x}) = f(x') + k' - t' r \in f(x') + K.$$
Because of the K-minimality of $\bar{x}$ we conclude $f(\bar{x}) = f(x')$ and thus $k' = t' r$. Due to the pointedness of the ordering cone K, $k' \in K$ and $t' r \in -K$ it follows $t' r = k' = 0_m$ in contradiction to $t' < 0$ and $r \neq 0_m$.

c) Assume $\bar{x}$ is not weakly K-minimal. Then there is a point $x' \in \Omega$ and a $k' \in \text{int}(K)$ with $f(\bar{x}) = f(x') + k'$. As $(\bar{t}, \bar{x})$ is a minimal solution and hence feasible for (SP(a, r)) there is a $\bar{k} \in K$ with $a + \bar{t} r - f(\bar{x}) = \bar{k}$. Because of $\bar{k} + k' \in \text{int}(K)$ there is a $\varepsilon > 0$ with $\bar{k} + k' - \varepsilon r \in \text{int}(K)$. Then we conclude from $a + \bar{t} r - f(x') = \bar{k} + k'$

2.2 Properties of the Pascoletti-Serafini Scalarization

$$a + (\bar{t} - \varepsilon)\, r - f(x') \in \text{int}(K).$$

Then the point $(\bar{t} - \varepsilon, x')$ is feasible for (SP(a, r)), too, with $\bar{t} - \varepsilon < \bar{t}$ in contradiction to $(\bar{t}, \bar{x})$ minimal. Using the same arguments we can show $\bar{k} \in \partial K$.

d) We assume $(0, \bar{x})$ is not a local minimal solution of (SP(a, r)). Then in any neighborhood $U = U_t \times U_x \subset \mathbb{R}^{n+1}$ of $(0, \bar{x})$ there exists a feasible point (t', x') with $t' < 0$ and a $k' \in K$ with $a + t' r - f(x') = k'$. With $a = f(\bar{x})$ we get

$$f(\bar{x}) = f(x') + k' - t'r.$$

Since $k' - t'r \in \text{int}(K)$ we have $f(\bar{x}) \in f(x') + \text{int}(K)$ and because the neighborhood U_x is arbitrarily chosen $\bar{x}$ cannot be locally weakly K-minimal.

e) With the same arguments as in the preceding proof we conclude again that if there exists a feasible point (t', x') with $t' < 0$ and x' in a neighborhood of $\bar{x}$ this leads to $f(\bar{x}) = f(x') + k' - t'r$ with $r \in K \setminus \{0_m\}$. Hence we have $f(\bar{x}) \in f(x') + K \setminus \{0_m\}$ in contradiction to $\bar{x}$ locally K-minimal.

f) Let $U = U_t \times U_x \subset \mathbb{R}^{n+1}$ be a neighborhood such that $(\bar{t}, \bar{x})$ is a local minimal solution of (SP(a, r)). Then there exists a $\bar{k} \in K$ with

$$a + \bar{t} r - f(\bar{x}) = \bar{k}. \tag{2.2}$$

We assume $\bar{x}$ is not a locally weakly K-minimal point of the multiobjective optimization problem (MOP). Then there exists no neighborhood $\bar{U}_x$ of $\bar{x}$ such that

$$f(\Omega \cap \bar{U}_x) \cap (f(\bar{x}) - \text{int}(K)) = \emptyset.$$

Hence for U_x there exists a point $x' \in \Omega \cap U_x$ with $f(x') \in f(\bar{x}) - \text{int}(K)$ and thus there is a $k' \in \text{int}(K)$ with $f(x') = f(\bar{x}) - k'$. Together with (2.2) we get

$$f(x') = a + \bar{t} r - \bar{k} - k'.$$

Because of $\bar{k} + k' \in \text{int}(K)$ there exists a $\varepsilon > 0$ with $\bar{t} - \varepsilon \in U_t$ and $\bar{k} + k' - \varepsilon r \in \text{int}(K)$. We conclude

$$f(x') = a + (\bar{t} - \varepsilon)\, r - (\bar{k} + k' - \varepsilon r)$$

and thus $(\bar{t} - \varepsilon, x') \in U$ is feasible for (SP(a, r)) with $\bar{t} - \varepsilon < \bar{t}$ in contradiction to $(\bar{t}, \bar{x})$ a local minimal solution. □

28   2 Scalarization Approaches

Remark 2.2. Note, that for the statement of Theorem 2.1,b) we need the pointedness of the ordering cone K. This is for instance not the case for the statement c).

Note also that it is not a consequence of Theorem 2.1,c) that we get always a weakly K-minimal point $\bar{x}$ by solving (SP(a,r)) for arbitrary parameters. It is possible that the problem (SP(a,r)) has no minimal solution at all as in the example shown in Fig. 2.3 for the case $m = 2$ and $K = \mathbb{R}^2_+$. There the minimal value of (SP(a,r)) is not bounded from below.

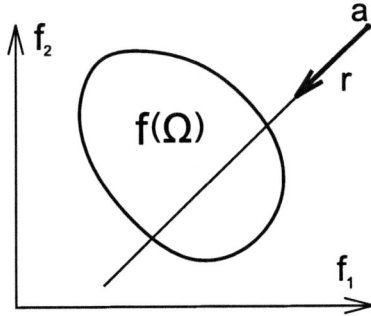

Fig. 2.3. For $K = \mathbb{R}^2_+$ there exists no minimal solution of problem (SP(a,r)).

For int$(K) = \emptyset$ we cannot apply the preceding theorem. However we can still consider the case of finding minimal points w.r.t. the relative algebraic interior (or intrinsic core, see also p.11) icr(K). For a closed convex cone $K \subset \mathbb{R}^m$ it is icr$(K) \neq \emptyset$ (compare [181, p.503]).

Theorem 2.3. *We consider the scalar optimization problem (SP(a,r)) to the multiobjective optimization problem (MOP) with $a \in \mathbb{R}^m$, $r \in L(K)$. Let $(\bar{t}, \bar{x})$ be a minimal solution, then $\bar{x}$ is minimal w.r.t. icr$(K) \cup \{0_m\}$.*

For the proof of this theorem we refer to [181]. If for a choice of parameters $a \in \mathbb{R}^m$ and $r \in \text{int}(K)$ the optimization problem (SP(a,r))

2.2 Properties of the Pascoletti-Serafini Scalarization

has no minimal solution, we can conclude under some additional assumptions that the related multiobjective optimization problem has no K-minimal solution at all. This is stated in the following theorem proven by Helbig in [104, Theorem 2.2].

Theorem 2.4. *Let $K \subset \mathbb{R}^m$ be a closed pointed convex cone with $int(K) \neq \emptyset$ and let the set $f(\Omega) + K$ be closed and convex. Assume $\mathcal{E}(f(\Omega), K) \neq \emptyset$. Then*

$$\{(a,r) \in \mathbb{R}^m \times int(K) \mid \Sigma(a,r) \neq \emptyset\} = \mathbb{R}^m \times int(K),$$

i. e. for any choice of parameters $(a,r) \in \mathbb{R}^m \times int(K)$ the scalar optimization problem (SP(a,r)) has feasible points.

Besides for all parameters $(a,r) \in \mathbb{R}^m \times int(K)$ there exists a minimal solution of (SP(a,r)).

As a direct consequence it follows:

Corollary 2.5. *Let $K \subset \mathbb{R}^m$ be a closed pointed convex cone with $int(K) \neq \emptyset$ and let the set $f(\Omega) + K$ be closed and convex. If there is a parameter $(a,r) \in \mathbb{R}^m \times int(K)$ such that (SP(a,r)) has no minimal solution then $\mathcal{E}(f(\Omega), K) = \emptyset$.*

Hence, if we solve the scalar problem (SP(a,r)) related to the multiobjective optimization problem (MOP) fulfilling the assumptions of Corollary 2.5 for an arbitrary choice of parameters $(a,r) \in \mathbb{R}^m \times int(K)$, then we either get a weakly K-minimal solution or we get the information that there are no efficient points of the problem (MOP). This property is not satisfied by all scalarization problems as e. g. not by the ε-constraint method as we will see later in Sect. 2.5.1.

For the special case of the natural ordering we have also the following similar theorem by Bernau ([15, Lemma 1.3]) not assuming the set $f(\Omega) + K$ to be closed and convex.

Theorem 2.6. *Let $\mathcal{M}_w(f(\Omega), \mathbb{R}^m_+) \neq \emptyset$. Then the objective function of the optimization problem (SP(a,r)) is bounded from below for arbitrary parameters $a \in \mathbb{R}^m$ and $r \in int(\mathbb{R}^m_+)$.*

Proof. Let $\bar{x} \in \mathcal{M}_w(f(\Omega), \mathbb{R}^m_+)$. We set

$$\bar{t} := \min_{1 \leq i \leq m} \frac{f_i(\bar{x}) - a_i}{r_i}. \tag{2.3}$$

Then $\bar{t} \leq \frac{1}{r_i}(f_i(\bar{x}) - a_i)$ for $i = 1, \ldots, m$. Next we assume there is a feasible point (t,x) of (SP(a,r)) with $t < \bar{t}$. Then $x \in \Omega$ and together with $r_i > 0$, $i = 1, \ldots, m$, and (2.3) it follows

$$f_i(x) \leq a_i + t\,r_i < a_i + \bar{t}\,r_i \leq f_i(\bar{x}) \qquad \text{for } i = 1,\ldots,m,$$

in contradiction to $\bar{x}$ weakly EP-minimal. Hence $\bar{t}$ is a lower bound for the objective function of the problem (SP(a,r)). □

We conclude:

Corollary 2.7. *If the objective value of the optimization problem (SP(a,r)) for $a \in \mathbb{R}^m$, $r \in int(\mathbb{R}_+^m)$ is not bounded from below then $\mathcal{M}(f(\Omega), \mathbb{R}_+^m) = \emptyset$, i. e. there exists no EP-minimal point of the related multiobjective optimization problem.*

An important property of the discussed scalarization approach is the possibility to generate all (weakly) K-minimal points of the multiobjective optimization problem (compare Theorem 2.1,a)). If $(\bar{t}, \bar{x})$ is a minimal solution of the scalar problem by Pascoletti-Serafini with $\bar{x}$ being weakly K-minimal but not K-minimal we have the following property for the points dominating the point $f(\bar{x})$ ([181, Theorem 3.3]).

Theorem 2.8. *If the point $(\bar{t}, \bar{x})$ is a minimal solution of (SP(a,r)) with $\bar{k} := a + \bar{t}\,r - f(\bar{x})$ and if there is a point $y = f(x) \in f(\Omega)$ dominating the point $f(\bar{x})$ w. r. t. the cone K, then the point $(\bar{t}, x)$ is also a minimal solution of (SP(a,r)) and there exists a $k \in \partial K$, $k \neq 0_m$, with $a + \bar{t}\,r - f(x) = \bar{k} + k$.*

From that we can immediately conclude:

Corollary 2.9. *If the point $(\bar{t}, \bar{x})$ is an image-unique minimal solution of the scalar problem (SP(a,r)) w. r. t. f, i. e. there is no other minimal solution (t, x) with $f(x) = f(\bar{x})$, then $\bar{x}$ is a K-minimal solution of the multiobjective optimization problem (MOP).*

Pascoletti and Serafini ([181, Theorem 3.7]) derive a criterion for checking whether a point is K-minimal or not.

Corollary 2.10. *A point $\bar{x}$ is a K-minimal solution of the multiobjective optimization problem (MOP) if*

i) there is some $\bar{t} \in \mathbb{R}$ so that $(\bar{t}, \bar{x})$ is a minimal solution of (SP(a,r)) for some parameters $a \in \mathbb{R}$ and $r \in int\,(K)$ and
ii) for $k := a + \bar{t}\,r - f(\bar{x})$ it is

$$((a + \bar{t}\,r) - \partial K) \cap (f(\bar{x}) - \partial K) \cap f(\Omega) = \{f(\bar{x})\}.$$

Hence if $(\bar{t}, \bar{x})$ is a minimal solution of $(SP(a,r))$ with $r \in \text{int}\,(K)$, then $\bar{x}$ is a weakly K-minimal solution and for checking if $\bar{x}$ is also K-minimal it is sufficient to test the points $((a+\bar{t}r) - \partial K) \cap (f(\bar{x}) - \partial K)$ of the set $f(\Omega)$.

2.3 Parameter Set Restriction for the Pascoletti-Serafini Scalarization

Our general aim is an approximation of the whole efficient set of the multiobjective optimization problem (MOP) by solving the problem $(SP(a,r))$ for several parameters. In Theorem 2.1,b) we have seen that we can find all K-minimal points for a constant parameter $r \in K \setminus \{0_m\}$ by varying the parameter $a \in \mathbb{R}^m$ only. In this section we show that we do not have to consider all parameters $a \in \mathbb{R}^m$. We can restrict the set from which we have to choose the parameter a such that we can still find all K-minimal points of the multiobjective optimization problem. We start by showing that it is sufficient to vary the parameter a on a hyperplane $H = \{y \in \mathbb{R}^m \mid b^\top y = \beta\}$ with $b \in \mathbb{R}^m \setminus \{0_m\}$, $\beta \in \mathbb{R}$.

Theorem 2.11. *Let $\bar{x}$ be K-minimal for (MOP) and define a hyperplane*

$$H = \{y \in \mathbb{R}^m \mid b^\top y = \beta\}$$

with $b \in \mathbb{R}^m \setminus \{0_m\}$ and $\beta \in \mathbb{R}$. Let $r \in K$ with $b^\top r \neq 0$ be arbitrarily given. Then there is a parameter $a \in H$ and some $\bar{t} \in \mathbb{R}$ so that $(\bar{t}, \bar{x})$ is a minimal solution of $(SP(a,r))$. This holds for instance for

$$\bar{t} = \frac{b^\top f(\bar{x}) - \beta}{b^\top r}$$

and

$$a = f(\bar{x}) - \bar{t}r.$$

Proof.   For

$$\bar{t} = \frac{b^\top f(\bar{x}) - \beta}{b^\top r} \quad \text{and} \quad a = f(\bar{x}) - \bar{t}r$$

we have $a \in H$ and the point $(\bar{t}, \bar{x})$ is feasible for $(SP(a,r))$. We assume that $(\bar{t}, \bar{x})$ is not a minimal solution of $(SP(a,r))$. Then there is a $t' \in \mathbb{R}$, $t' < \bar{t}$, and points $x' \in \Omega$ and $k' \in K$ with

$$a + t'r - f(x') = k'.$$

With the definition of a it follows

$$f(\bar{x}) - \bar{t}\,r + t'\,r - f(x') = k'.$$

Hence

$$f(\bar{x}) = f(x') + \underbrace{k' + (\bar{t}-t')}_{>0}\underbrace{r}_{\in K}$$

and because of the convexity of the cone K

$$f(\bar{x}) \in f(x') + K. \qquad (2.4)$$

As the cone K is pointed and because $r \neq 0_m$ it is

$$\underbrace{k' + (\bar{t}-t')}_{>0} r \neq 0_m$$

and thus $f(\bar{x}) \neq f(x')$. With (2.4) we conclude $f(\bar{x}) \in f(x') + K \setminus \{0_m\}$ for $x' \in \Omega$ in contradiction to $\bar{x}$ K-minimal. $\square$

In the following we give a stricter restriction of the set from which we have to choose the parameter a such that we are still able to find all K-minimal points. We first consider the bicriteria case before we come to the more general case of an arbitrary multiobjective optimization problem.

2.3.1 Bicriteria Case

In this section we only consider biobjective problems, i.e. let $m = 2$ except when otherwise stated. In the preceding theorem we have seen that it is sufficient to choose the parameter $r \in K \setminus \{0_m\}$ constant and to vary the parameter a only in a hyperplane

$$H = \{y \in \mathbb{R}^2 \mid b_1 y_1 + b_2 y_2 = \beta\}$$

(here a line) with $b = (b_1, b_2) \in \mathbb{R}^2$, $b^\top r \neq 0$ and $\beta \in \mathbb{R}$. For example we can choose $b = r$ and $\beta = 0$, then $b^\top r = r^\top r = r_1^2 + r_2^2 \neq 0$ for $r \neq 0_2$.

In the bicriteria case we have the property that any closed pointed ordering cone in $\mathbb{R}^2$ is polyhedral (see Lemma 1.20). By using this property we can show that it is sufficient to consider only a subset H^a of the hyperplane H. In the following we assume $r \in K \setminus \{0_2\}$.

2.3 Parameter Set Restriction

We first consider the case that the ordering cone K has a nonempty interior and is thus given by

$$K = \left\{ y \in \mathbb{R}^2 \mid l^{1\top} y \geq 0,\ l^{2\top} y \geq 0 \right\} \tag{2.5}$$

with $l^1,\ l^2 \in \mathbb{R}^2 \setminus \{0_2\}$, $l^1,\ l^2$ linearly independent. Then the interior of the cone is $\mathrm{int}(K) = \{ y \in \mathbb{R}^2 \mid l^{1\top} y > 0,\ l^{2\top} y > 0 \}$. Assuming the set $f(\Omega)$ to be compact there exists a minimal solution $\bar{x}^1$ of the scalar-valued problem

$$\min_{x \in \Omega} l^{1\top} f(x) \tag{2.6}$$

and a minimal solution $\bar{x}^2$ of the scalar-valued problem

$$\min_{x \in \Omega} l^{2\top} f(x). \tag{2.7}$$

These minimal solutions are also weakly K-minimal solutions of the multiobjective optimization problem (MOP) with the vector-valued objective function $f \colon \mathbb{R}^n \to \mathbb{R}^2$ as it is shown in the following lemma. This result can be generalized to the case with more than two objectives, too, for a finitely generated ordering cone $K \subset \mathbb{R}^m$. Besides, as the minimal solutions of (2.6) and (2.7) are weakly K-minimal, there are also parameters a and r such that these points are minimal solutions of the scalar problem (SP(a, r)).

Lemma 2.12. *We consider the multiobjective optimization problem (MOP) for $m \in \mathbb{N}$, $m \geq 2$. Let $K \subset \mathbb{R}^m$ be a finitely generated cone with nonempty interior given by*

$$K = \{ y \in \mathbb{R}^m \mid l^{i\top} y \geq 0,\ i = 1, \ldots, s \}$$

($s \in \mathbb{N}$). Let $\bar{x}^j$ be a minimal solution of

$$\min_{x \in \Omega} l^{j\top} f(x) \tag{2.8}$$

for a $j \in \{1, \ldots, s\}$. Then $\bar{x}^j$ is weakly K-minimal.

If we consider now the scalarization problem (SP(a, r)) with parameters $r \in K$, with $l^{j\top} r > 0$ (e. g. satisfied for $r \in \mathrm{int}(K)$), and $a := \bar{a}^j$ given by

$$\bar{a}^j := f(\bar{x}^j) - \bar{t}^j r \quad \text{with} \quad \bar{t}^j := \frac{b^\top f(\bar{x}^j) - \beta}{b^\top r}$$

for $b \in \mathbb{R}^m$, $b^\top r \neq 0$, $\beta \in \mathbb{R}$, then $(\bar{t}^j, \bar{x}^j)$ is a minimal solution of (SP($\bar{a}^j, r$)).

Proof. We first show $\bar{x}^j \in \mathcal{M}_w(f(\Omega), K)$. For that we assume that the point $\bar{x}^j$ is not weakly K-minimal. Then there is a point $x \in \Omega$ with
$$f(\bar{x}^j) \in f(x) + \text{int}(K).$$
Then it follows $l^{j\top}(f(\bar{x}^j) - f(x)) > 0$ and hence $l^{j\top} f(\bar{x}^j) > l^{j\top} f(x)$ in contradiction to $\bar{x}^j$ a minimal solution of (2.8).

Next we show that $(\bar{t}^j, \bar{x}^j)$ is a minimal solution of $(\text{SP}(\bar{a}^j, r))$. Because of
$$\bar{a}^j + \bar{t}^j r - f(\bar{x}^j) = 0_m$$
the point $(\bar{t}^j, \bar{x}^j)$ is a feasible point. We now assume that this point is not a minimal solution. Then there exists a feasible point (t', x') with $t' < \bar{t}^j$. Because of the feasibility of (t', x') for $(\text{SP}(\bar{a}^j, r))$ it holds
$$\bar{a}^j + t' r - f(x') \in K.$$
Together with the definition of $\bar{a}^j$ we conclude
$$f(\bar{x}^j) - \bar{t}^j r + t' r - f(x') \in K.$$
Then
$$l^{j\top}\left(f(\bar{x}^j) + (t' - \bar{t}^j) r - f(x')\right) \geq 0$$
and thus
$$l^{j\top} f(\bar{x}^j) \geq l^{j\top} f(x') + \underbrace{(\bar{t}^j - t')}_{>0} \underbrace{l^{j\top} r}_{>0}$$
$$> l^{j\top} f(x')$$
in contradiction to $\bar{x}^j$ a minimal solution of (2.8). □

The second result of this lemma is no longer true for arbitrary $r \in \partial K = K \setminus \text{int}(K)$ with $l^{j\top} r = 0$ as it is demonstrated in the following example.

Example 2.13. We consider the bicriteria optimization problem
$$\min \begin{pmatrix} x_1 \\ x_2 \end{pmatrix}$$
subject to the constraints
$$1 \leq x_1 \leq 3,$$
$$1 \leq x_2 \leq 3,$$
$$x \in \mathbb{R}^2$$

2.3 Parameter Set Restriction   35

w. r. t. the ordering cone $K = \mathbb{R}_+^2 = \{y \in \mathbb{R}^2 \mid (1,0)\, y \geq 0,\ (0,1)\, y \geq 0\}$.
Then for $l^1 = (1,0)^\top$ the point $\bar{x}^1 = (1,2)$ is a minimal solution of (2.8) for $j = 1$ and also a weakly EP-minimal solution of the bicriteria problem. For the hyperplane (here a line) $H := \{y \in \mathbb{R}^2 \mid (0,1)\, y = 0\}$, i. e. $b = (0,1)^\top$, $\beta = 0$, and the parameter $r := (0,1)^\top$ we get according to Lemma 2.12 $\bar{t} = 2$ and $\bar{a}^1 = (1,0)^\top$. However the point $(\bar{t}^1, \bar{x}^1) = (2,1,2)$ is not a minimal solution of $(\mathrm{SP}(\bar{a}^1, r))$. The point $(1,1,1)$ is the unique minimal solution of $(\mathrm{SP}(\bar{a}^1, r))$. For $r = (0,1)^\top$ there is no parameter $a \in H$ at all such that there exists a t with $(t, \bar{x}^1)$ a minimal solution of $(\mathrm{SP}(a, r))$.

Remark 2.14. If we extend the assumptions in Lemma 2.12 by the assumption that the minimal solution of (2.8) is unique then we can drop the condition $l^{j\top} r > 0$ and the result is also valid for r with $l^{j\top} r \geq 0$ what is already fulfilled for $r \in K$.

We now concentrate again on the bicriteria case for which we get further results.

Lemma 2.15. *We consider the multiobjective optimization problem (MOP) for $m = 2$ with the ordering cone $K \subset \mathbb{R}^2$ given by*

$$K = \{y \in \mathbb{R}^2 \mid l^{i\top} y \geq 0,\ i = 1, 2\}.$$

Let $\bar{x}^1$ be a minimal solution of (2.6) and let $\bar{x}^2$ be a minimal solution of (2.7). Then for all $x \in \mathcal{M}(f(\Omega), K)$ it is

$$l^{1\top} f(\bar{x}^1) \leq l^{1\top} f(x) \leq l^{1\top} f(\bar{x}^2)$$

and

$$l^{2\top} f(\bar{x}^2) \leq l^{2\top} f(x) \leq l^{2\top} f(\bar{x}^1).$$

Proof. As $\bar{x}^1$ and $\bar{x}^2$ are minimal solutions of (2.6) and (2.7) we have of course for all $x \in \mathcal{M}(f(\Omega), K) \subset \Omega$

$$l^{1\top} f(x) \geq l^{1\top} f(\bar{x}^1) \text{ and } l^{2\top} f(x) \geq l^{2\top} f(\bar{x}^2).$$

Let us now suppose that there is a point $x \in \mathcal{M}(f(\Omega), K)$ with $l^{2\top} f(x) > l^{2\top} f(\bar{x}^1)$, i.e. with $l^{2\top} (f(x) - f(\bar{x}^1)) > 0$. Together with $l^{1\top} (f(x) - f(\bar{x}^1)) \geq 0$ we get $f(x) - f(\bar{x}^1) \in K \setminus \{0_2\}$ in contradiction to x K-minimal. Thus we have shown that $l^{2\top} f(x) \leq l^{2\top} f(\bar{x}^1)$ has to be true. The same for $l^{1\top} f(x) \leq l^{1\top} f(\bar{x}^2)$. □

We conclude:

Lemma 2.16. *Let the assumptions of Lemma 2.15 hold. If the efficient set $\mathcal{E}(f(\Omega), K)$ consists of more than one point, what is generally the case, then*

$$l^{1\top} f(\bar{x}^1) < l^{1\top} f(\bar{x}^2)$$

and

$$l^{2\top} f(\bar{x}^2) < l^{2\top} f(\bar{x}^1).$$

Proof. As $\bar{x}^1$ is a minimal solution of (2.6) we already have $l^{1\top} f(\bar{x}^1) \leq l^{1\top} f(\bar{x}^2)$. We assume now $l^{1\top} f(\bar{x}^1) = l^{1\top} f(\bar{x}^2)$. Applying Lemma 2.15 we get $l^{1\top} f(x) = l^{1\top} f(\bar{x}^2)$ for all $x \in \mathcal{M}(f(\Omega), K)$ and thus $l^{1\top}(f(x) - f(\bar{x}^2)) = 0$. Further, according to Lemma 2.15 it is $l^{2\top} f(x) \geq l^{2\top} f(\bar{x}^2)$ and hence $l^{2\top}(f(x) - f(\bar{x}^2)) \geq 0$ for all $x \in \mathcal{M}(f(\Omega), K)$.

Summarizing this results in $f(x) - f(\bar{x}^2) \in K$. As x is K-minimal we conclude $f(x) = f(\bar{x}^2)$ for all $x \in \mathcal{M}(f(\Omega), K)$ and thus $\mathcal{E}(f(\Omega), K) = \{f(\bar{x}^2)\}$.

Analogously $l^{2\top} f(\bar{x}^1) = l^{2\top} f(\bar{x}^2)$ implies $\mathcal{E}(f(\Omega), K) = \{f(\bar{x}^1)\}$. □

We project the points $f(\bar{x}^1)$ and $f(\bar{x}^2)$ in direction r onto the line H (compare Fig. 2.4 for $l^1 = (1,0)$ and $l^2 = (0,1)$, i.e. $K = \mathbb{R}_+^2$). The projection points $\bar{a}^1 \in H = \{y \in \mathbb{R}^2 \mid b^\top y = \beta\}$ and $\bar{a}^2 \in H$ are given by

$$\bar{a}^i := f(\bar{x}^i) - \bar{t}^i r \quad \text{with} \quad \bar{t}^i := \frac{b^\top f(\bar{x}^i) - \beta}{b^\top r}, \qquad i = 1, 2. \qquad (2.9)$$

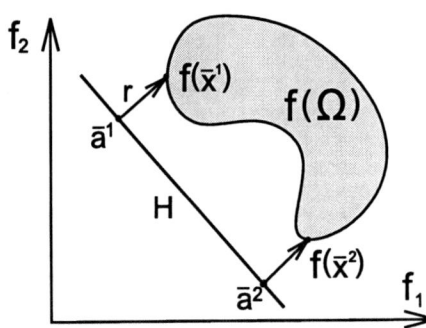

Fig. 2.4. Projection of the points $f(\bar{x}^1)$ and $f(\bar{x}^2)$ in direction r onto H.

2.3 Parameter Set Restriction

We show that it is sufficient to consider parameters $a \in H^a$ with the set H^a given by

$$H^a = \{y \in H \mid y = \lambda \bar{a}^1 + (1-\lambda)\bar{a}^2, \ \lambda \in [0,1]\}, \tag{2.10}$$

i.e. it is sufficient to consider parameters on the line H between the points $\bar{a}^1$ and $\bar{a}^2$.

Theorem 2.17. *We consider the multiobjective optimization problem (MOP) with $m=2$ and K as in (2.5). Further let $\bar{a}^1$ and $\bar{a}^2$ be given as in (2.9) with $\bar{x}^1$ and $\bar{x}^2$ minimal solutions of (2.6) and (2.7) respectively. Then we have for the set H^a as defined in (2.10) $H^a \subset H$ and for any K-minimal solution $\bar{x}$ of (MOP) there exists a parameter $a \in H^a$ and some $\bar{t} \in \mathbb{R}$ so that $(\bar{t}, \bar{x})$ is a minimal solution of (SP(a,r)).*

Proof. Because of $\bar{a}^1, \bar{a}^2 \in H$ it is $H^a \subset H$. According to Theorem 2.11 we already have that for any $\bar{x} \in \mathcal{M}(f(\Omega), K)$ there exists a parameter $a \in H$ and a $\bar{t} \in \mathbb{R}$ given by

$$\bar{t} = \frac{b^\top f(\bar{x}) - \beta}{b^\top r} \quad \text{and} \quad a = f(\bar{x}) - \bar{t} r$$

so that $(\bar{t}, \bar{x})$ is a minimal solution of (SP(a,r)). Hence it is sufficient to show that the parameter a lies on the line segment between the points $\bar{a}^1$ and $\bar{a}^2$, i.e. that $a = \lambda \bar{a}^1 + (1-\lambda)\bar{a}^2$ for a $\lambda \in [0,1]$. Using the definitions of a, $\bar{a}^1$ and $\bar{a}^2$ the equation $a = \lambda \bar{a}^1 + (1-\lambda)\bar{a}^2$ is equivalent to

$$f(\bar{x}) - \bar{t} r = \lambda \left(f(\bar{x}^1) - \bar{t}^1 r\right) + (1-\lambda)\left(f(\bar{x}^2) - \bar{t}^2 r\right). \tag{2.11}$$

If the efficient set of the multiobjective optimization problem consists of one point only and thus of $f(\bar{x}^1)$ or $f(\bar{x}^2)$ only, then (2.11) is satisfied for $\lambda = 1$ or $\lambda = 0$ respectively. Otherwise we have according to the Lemma 2.16

$$l^{1\top} f(\bar{x}^1) < l^{1\top} f(\bar{x}^2) \tag{2.12}$$

and

$$l^{2\top} f(\bar{x}^2) < l^{2\top} f(\bar{x}^1). \tag{2.13}$$

We reformulate the equation (2.11) as

$$f(\bar{x}) = \lambda f(\bar{x}^1) + (1-\lambda) f(\bar{x}^2) + \left(\bar{t} - \lambda \bar{t}^1 - (1-\lambda)\bar{t}^2\right) r. \tag{2.14}$$

Then we can do a case differentiation for

$$\bar{t} - \lambda \bar{t}^1 - (1-\lambda)\bar{t}^2 = \frac{1}{b^\top r}(b^\top(f(\bar{x}) - \lambda f(\bar{x}^1) - (1-\lambda)f(\bar{x}^2))) \geq 0$$

and $\bar{t} - \lambda \bar{t}^1 - (1-\lambda)\bar{t}^2 < 0$ respectively.

For the case $\bar{t} - \lambda \bar{t}^1 - (1-\lambda)\bar{t}^2 \geq 0$ we start by assuming that (2.14) is only satisfied for $\lambda < 0$. By applying the linear map l^1 to (2.14) we get because of $r \in K$ and together with (2.12)

$$l^{1\top} f(\bar{x}) = \lambda l^{1\top} f(\bar{x}^1) + (1-\lambda) l^{1\top} f(\bar{x}^2) + \underbrace{(\bar{t} - \lambda \bar{t}^1 - (1-\lambda)\bar{t}^2)}_{\geq 0} \underbrace{l^{1\top} r}_{\geq 0}$$

$$\geq \underbrace{\lambda \underbrace{l^{1\top} f(\bar{x}^1)}_{<l^{1\top}f(\bar{x}^2)}}_{<0} + (1-\lambda) l^{1\top} f(\bar{x}^2)$$

$$> \lambda l^{1\top} f(\bar{x}^2) + (1-\lambda) l^{1\top} f(\bar{x}^2)$$

$$= l^{1\top} f(\bar{x}^2)$$

in contradiction to Lemma 2.15.

Now we suppose (2.14) is only satisfied for $\lambda > 1$. By applying the linear map l^2 to (2.14) and together with (2.13) we conclude

$$l^{2\top} f(\bar{x}) \geq \lambda l^{2\top} f(\bar{x}^1) + \underbrace{(1-\lambda)}_{<0} \underbrace{l^{2\top} f(\bar{x}^2)}_{<l^{2\top}f(\bar{x}^1)}$$

$$> l^{2\top} f(\bar{x}^1)$$

in contradiction to Lemma 2.15. Thus we have shown that for the case $\bar{t} - \lambda \bar{t}^1 - (1-\lambda)\bar{t}^2 \geq 0$ it is $\lambda \in [0,1]$.

For the case $\bar{t} - \lambda \bar{t}^1 - (1-\lambda)\bar{t}^2 < 0$ one can show analogously that $\lambda \in [0,1]$. □

Remark 2.18. An even more strict restriction of the parameter set H^a is possible by minimizing in (2.6) and (2.7) over the set $\mathcal{M}(f(\Omega), K)$ instead of over Ω. Then we still have that for any $\bar{x} \in \mathcal{M}(f(\Omega), K)$ there exists a parameter $a \in H^a$ and some scalar $\bar{t} \in \mathbb{R}$ so that $(\bar{t}, \bar{x})$ is a minimal solution of (SP(a, r)). However the set $\mathcal{M}(f(\Omega), K)$ is generally not known and thus we cannot optimize over the set of minimal solutions of the multiobjective optimization problem.

Next we come to the case that the ordering cone K has an empty interior, i.e. is given by $K = \{\lambda k \mid \lambda \geq 0\}$ for a $k \in \mathbb{R}^2 \setminus \{0_2\}$. Then the scalar optimization problem (SP(a,r)) with $r \in K \setminus \{0_2\}$, i.e. $r = \lambda^r k$ for a $\lambda^r > 0$, can be formulated as

2.3 Parameter Set Restriction

$$\min t$$

subject to the constraints

$$a + (t\lambda^r - \lambda)k = f(x),$$
$$t \in \mathbb{R}, \ x \in \Omega, \ \lambda \geq 0$$

(2.15)

by introducing an additional variable $\lambda \in \mathbb{R}$. If a point $(\bar{t}, \bar{x}, \bar{\lambda})$ is a minimal solution of (2.15) we always have $\bar{\lambda} = 0$: suppose $(\bar{t}, \bar{x}, \bar{\lambda})$ is a minimal solution of (2.15) with $\bar{\lambda} > 0$. Then the point $(\bar{t} - \frac{\bar{\lambda}}{\lambda^r}, \bar{x}, 0)$ is also feasible for (2.15) with $\bar{t} - \frac{\bar{\lambda}}{\lambda^r} < \bar{t}$ in contradiction to $(\bar{t}, \bar{x}, \bar{\lambda})$ a minimal solution. Thus we can consider the problem

$$\min t$$

subject to the constraints

$$a + tr = f(x),$$
$$t \in \mathbb{R}, \ x \in \Omega,$$

(2.16)

instead of (2.15).

To determine the set H^a as a subset of the set H it is sufficient to project the set $f(\Omega)$ in direction r onto the hyperplane H (see Fig. 2.5). If we have $l \in \mathbb{R}^2 \setminus \{0_2\}$ with $l^\top r = 0$ then we can determine the set H^a as described in Theorem 2.17 by solving the problems (2.6) and (2.7) with $l^1 := l$ and $l^2 := -l$.

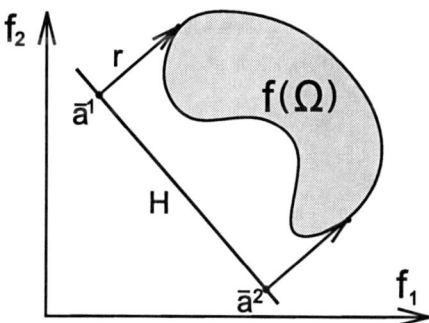

Fig. 2.5. Projection of the set $f(\Omega)$ in direction r on the hyperplane H.

2.3.2 General Case

Compared to the case with only two objective functions it is more difficult to restrict the parameter set in the case of three and more criteria. For example not any closed pointed convex cone in $\mathbb{R}^m$, $m \geq 3$, is polyhedral unlike it is in the case for $m = 2$ (Lemma 1.20). The cone of Example 1.21 which represents the Löwner partial ordering is non-polyhedral and thus not finitely generated. However even for polyhedral cones the results from Sect. 2.3.1 cannot be generalized to $\mathbb{R}^3$. A finitely generated cone $K \subset \mathbb{R}^3$ given by

$$K = \{y \in \mathbb{R}^3 \mid l^{i\top} y \geq 0, \ i = 1, \ldots, s\}$$

with $l^i \in \mathbb{R}^3 \setminus \{0_3\}$, $i = 1, \ldots, s$, $s \in \mathbb{N}$, does not need to be generated by $s = m = 3$ vectors l^i only. Instead it is possible that $s > m$, as it is shown in Example 1.19.

Even if the ordering cone K is finitely generated by three vectors, i.e. $s = 3$ as it is the case for the ordering cone $K = \mathbb{R}^3_+$ inducing the natural ordering, we cannot generalize the results gained in the preceding section for determining the set H^a. This is illustrated with the following example.

Example 2.19. We consider the objective function $f \colon \mathbb{R}^3 \to \mathbb{R}^3$ with $f(x) := x$ for all $x \in \mathbb{R}^3$ and the constraint set $\Omega \subset \mathbb{R}^3$ defined by

$$\Omega := \{x = (x_1, x_2, x_3) \in \mathbb{R}^3 \mid x_1^2 + x_2^2 + x_3^2 \leq 1\}$$

which equals the unit ball in $\mathbb{R}^3$. We assume that the ordering is induced by the cone $K := \mathbb{R}^3_+$ which is finitely generated by

$$l^1 := (1,0,0)^\top, \ l^2 := (0,1,0)^\top, \text{ and } l^3 := (0,0,1)^\top.$$

Thus $K = \{y \in \mathbb{R}^3 \mid l^{i\top} y \geq 0, \ i = 1, 2, 3\}$. The tricriteria optimization problem

$$\min_{x \in \Omega} f(x)$$

has the solution set

$$\mathcal{M}(f(\Omega), \mathbb{R}^3_+) = \{x \in \mathbb{R}^3 \mid x_1^2 + x_2^2 + x_3^2 = 1, \ x_i \leq 0, \ i = 1, 2, 3\}.$$

By solving the three scalar optimization problems

$$\min_{x \in \Omega} l^{i\top} f(x)$$

($i = 1, 2, 3$) corresponding to the problems (2.6) and (2.7) we get the three minimal solutions

$$\bar{x}^1 = (-1, 0, 0)^\top, \ \bar{x}^2 = (0, -1, 0)^\top, \ \text{and} \ \bar{x}^3 = (0, 0, -1)^\top.$$

Further define a hyperplane by

$$H := \{y \in \mathbb{R}^3 \mid (-1, -1, -1) \cdot y = 1\}$$

with $b = (-1, -1, -1)^\top, \beta = 1$. Then it is $f(\bar{x}^i) = \bar{x}^i \in H$ for $i = 1, 2, 3$.
For $r := (1, 1, 1)^\top$ the points $\bar{a}^i \in H$, $i = 1, 2, 3$, gained analogously to the points in (2.9), are

$$\bar{a}^1 = (-1, 0, 0)^\top, \ \bar{a}^2 = (0, -1, 0)^\top, \ \text{and} \ \bar{a}^3 = (0, 0, -1)^\top.$$

By defining the set H^a as the set of all convex combinations of the points $\bar{a}^i$, $i = 1, 2, 3$, as it is done in the bicriteria case, we get

$$H^a := \left\{ y \in \mathbb{R}^3 \ \Big| \ y = \sum_{i=1}^{3} \lambda_i \bar{a}^i, \ \lambda_i \geq 0, \ i = 1, 2, 3, \ \sum_{i=1}^{3} \lambda_i = 1 \right\}.$$

Then it is no longer true that to any K-minimal point $\bar{x}$ of the multiobjective optimization problem there is a parameter $\bar{a} \in H^a$ and a $\bar{t} \in \mathbb{R}$ so that $(\bar{t}, \bar{x})$ is a minimal solution of $(\mathrm{SP}(\bar{a}, r))$. For example the point $\bar{x} = (-1/\sqrt{2}, -1/\sqrt{2}, 0)^\top$ is EP-minimal but there is no parameter $\bar{a} \in H^a$ such that we get the point $\bar{x}$ by solving $(\mathrm{SP}(\bar{a}, r))$. For $\bar{a} = -1/(3\sqrt{2}) \cdot (1 + \sqrt{2}, 1 + \sqrt{2}, \sqrt{2} - 2)^\top$ and $\bar{t} = (1 - \sqrt{2})/3$ the point $(\bar{t}, \bar{x})$ is a minimal solution of $(\mathrm{SP}(\bar{a}, r))$, but it is $\bar{a} \notin H^a$.

Das and Dennis are confronted with the same problem during their examinations of the normal boundary intersection method in [40, pp.635f].

Due to these difficulties we determine a weaker restriction of the set H for the parameter a by projecting the image set $f(\Omega)$ in direction r onto the set H. Thus we determine the set

$$\tilde{H} := \{y \in H \mid y + t\,r = f(x), \ t \in \mathbb{R}, \ x \in \Omega\} \subset H \quad (2.17)$$

(see Fig. 2.6 for $m = 3$).

We would again (see Remark 2.18) get a better result in the sense of a stronger restriction of the set H by projecting the efficient points of the set $f(\Omega)$, i.e. the set $\mathcal{E}(f(\Omega), K)$, only onto the hyperplane H, i.e.

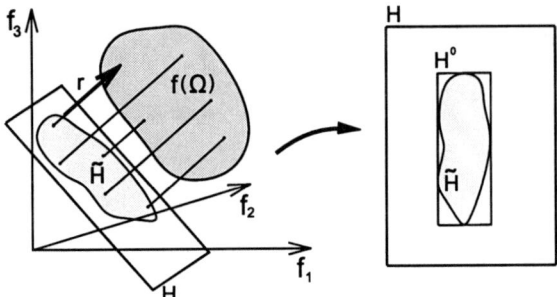

Fig. 2.6. Determination of the sets $\tilde{H}$ and H^0.

by determining $\{y \in H \mid y + tr = f(x),\ t \in \mathbb{R},\ x \in \mathcal{M}(f(\Omega), K)\} \subset H$. However the set of K-minimal points is generally not known. It is the aim of our method to approximate this set.

The set $\tilde{H} \subset H$ has in general an irregular boundary and is therefore not suitable for a systematic procedure. Hence we embed the set $\tilde{H}$ in a $(m-1)$-dimensional cuboid $H^0 \subset \mathbb{R}^m$ which is chosen as minimal as possible. For calculating the set H^0 we first determine $m-1$ vectors $v^1, \ldots, v^{m-1}$, which span the hyperplane H with $\tilde{H} \subset H$ and which are orthogonal and normalized by one, i.e.

$$v^{i\top} v^j = \begin{cases} 0, & \text{for } i \neq j, \quad i,j \in \{1,\ldots,m-1\}, \\ 1, & \text{for } i = j, \quad i,j \in \{1,\ldots,m-1\}. \end{cases} \quad (2.18)$$

These vectors form an orthonormal basis of the smallest subspace of $\mathbb{R}^m$ containing H. We have the condition $v^i \in H$, $i = 1, \ldots, m-1$, i.e.

$$b^\top v^i = \beta, \qquad i = 1, \ldots, m-1. \quad (2.19)$$

For example for $m = 3$ we can choose v^1 and v^2 dependent on $b = (b_1, b_2, b_3)^\top$ as follows

$$\tilde{v}^1 := \begin{cases} (\frac{b_1}{b_3}, -\frac{b_3}{b_2} - \frac{b_1^2}{b_2 b_3}, 1)^\top & \text{if } b_2 \neq 0,\ b_3 \neq 0, \\ (b_3, 0, -b_1)^\top & \text{if } b_2 = 0,\ b_3 \neq 0, \\ (0, 0, -b_1)^\top & \text{if } b_1 \neq 0,\ b_3 = 0, \\ (1, 0, 0)^\top & \text{if } b_1 = 0,\ b_3 = 0, \end{cases} \quad \text{and } v^1 := \frac{\tilde{v}^1}{\|\tilde{v}^1\|_2}$$

as well as

2.3 Parameter Set Restriction

$$\tilde{v}^2 := \begin{cases} (b_3, 0, -b_1)^\top & \text{if } b_2 \neq 0,\ b_3 \neq 0, \\ (0, 1, 0)^\top & \text{if } b_2 = 0,\ b_3 \neq 0, \\ (-\frac{b_2}{b_1}, 1, 0)^\top & \text{if } b_1 \neq 0,\ b_3 = 0, \\ (0, 0, 1)^\top & \text{if } b_1 = 0,\ b_3 = 0, \end{cases} \quad \text{and hence } v^2 := \frac{\tilde{v}^2}{\|\tilde{v}^2\|_2}.$$

This leads to the representation

$$H = \{y \in \mathbb{R}^m \mid y = \sum_{i=1}^{m-1} s_i v^i,\ s \in \mathbb{R}^{m-1}\}. \tag{2.20}$$

of the hyperplane H. Then, for the set $\tilde{H}$ as in (2.17), we can determine the searched cuboid by solving the following $2(m-1)$ scalar-valued optimization problems

$$\begin{aligned} & \min s_j \\ & \text{subject to the constraints} \\ & \sum_{i=1}^{m-1} s_i v^i + t\,r = f(x), \\ & t \in \mathbb{R}, \\ & x \in \Omega, \\ & s \in \mathbb{R}^{m-1} \end{aligned} \tag{2.21}$$

for $j \in \{1, \ldots, m-1\}$ with minimal solution $(t^{\min,j}, x^{\min,j}, s^{\min,j})$ and minimal value $s_j^{\min,j}$ and

$$\begin{aligned} & \min -s_j \\ & \text{subject to the constraints} \\ & \sum_{i=1}^{m-1} s_i v^i + t\,r = f(x), \\ & t \in \mathbb{R}, \\ & x \in \Omega, \\ & s \in \mathbb{R}^{m-1} \end{aligned} \tag{2.22}$$

for $j \in \{1, \ldots, m-1\}$ with minimal solution $(t^{\max,j}, x^{\max,j}, s^{\max,j})$ and minimal value $-s_j^{\max,j}$. We get

$$H^0 := \left\{ y \in \mathbb{R}^m \ \middle| \ y = \sum_{i=1}^{m-1} s_i v^i,\ s_i \in [s_i^{\min,i}, s_i^{\max,i}],\ i = 1, \ldots, m-1 \right\}$$

with $\tilde{H} \subset H^0$. This is a suitable restriction of the parameter set H as the following lemma shows.

Lemma 2.20. *Let $\bar{x}$ be a K-minimal solution of the multiobjective optimization problem (MOP). Let $r \in K \setminus \{0_m\}$. Then there is a parameter $\bar{a} \in H^0$ and some $\bar{t} \in \mathbb{R}$ so that $(\bar{t}, \bar{x})$ is a minimal solution of $(SP(\bar{a}, r))$.*

Proof. According to Theorem 2.11 the point $(\bar{t}, \bar{x})$ with

$$\bar{t} := \frac{b^\top f(\bar{x}) - \beta}{b^\top r}$$

is a minimal solution of $(SP(\bar{a}, r))$ for $\bar{a} := f(\bar{x}) - \bar{t} r \in H$. Because of $H^0 \subset H$ it suffices to show $\bar{a} \in H^0$. As $\bar{a} \in H$ there is according to the representation in (2.20) a point $\bar{s} \in \mathbb{R}^{m-1}$ with

$$\bar{a} = \sum_{i=1}^{m-1} \bar{s}_i v^i.$$

Because of $\bar{a} + \bar{t} r = f(\bar{x})$ the point $(\bar{t}, \bar{x}, \bar{s})$ is feasible for the optimization problems (2.21) and (2.22). Thus it is $s_i^{\min,i} \leq \bar{s}_i \leq s_i^{\max,i}$ for $i = 1, \ldots, m-1$ and it follows $\bar{a} \in H^0$. □

Hence we can also restrict the parameter set for the case of more than two objectives and arbitrary ordering cones K.

2.4 Modified Pascoletti-Serafini Scalarization

For theoretical reasons we are also interested in the following modification of the Pascoletti-Serafini problem named $(\overline{SP}(a, r))$ which is given by

$$\begin{aligned} & \min t \\ & \text{subject to the constraints} \\ & a + t r - f(x) = 0_m, \\ & t \in \mathbb{R}, x \in \Omega. \end{aligned} \quad (2.23)$$

Here the inequality constraint $a + t r - f(x) \in K$ is replaced by an equality constraint. For the connection between the problem $(SP(a, r))$ and the problem $(\overline{SP}(a, r))$ the following theorem is important.

2.4 Modified Pascoletti-Serafini Scalarization

Theorem 2.21. *Let a hyperplane* $H = \{y \in \mathbb{R}^m \mid b^\top y = \beta\}$ *with* $b \in \mathbb{R}^m \setminus \{0_m\}$ *and* $\beta \in \mathbb{R}$ *be given. Let* $(\bar{t}, \bar{x})$ *be a minimal solution of the scalar optimization problem (SP(a,r)) for the parameters* $a \in \mathbb{R}^m$ *and* $r \in \mathbb{R}^m$ *with* $b^\top r \neq 0$. *Hence there is a* $\bar{k} \in K$ *with*

$$a + \bar{t}r - f(\bar{x}) = \bar{k}.$$

Then there is a parameter $a' \in H$ *and some* $t' \in \mathbb{R}$ *so that* $(t', \bar{x})$ *is a minimal solution of (SP(a',r)) with*

$$a' + t'r - f(\bar{x}) = 0_m.$$

Proof. We set

$$t' := \frac{b^\top f(\bar{x}) - \beta}{b^\top r}$$

and

$$a' := a + (\bar{t} - t')r - \bar{k} = f(\bar{x}) - t'r.$$

Then $a' \in H$ and $a' + t'r - f(\bar{x}) = 0_m$. The point $(t', \bar{x})$ is feasible for (SP(a',r)) and it is also a minimal solution, because otherwise there exists a feasible point $(\hat{t}, \hat{x})$ of (SP(a',r)) with $\hat{t} < t'$, $\hat{x} \in \Omega$, and some $\hat{k} \in K$ with

$$a' + \hat{t}r - f(\hat{x}) = \hat{k}.$$

Together with the definition of a' we conclude

$$a + (\bar{t} - t' + \hat{t})r - f(\hat{x}) = \hat{k} + \bar{k} \in K.$$

Hence $(\bar{t} - t' + \hat{t}, \hat{x})$ is feasible for (SP(a,r)) with $\bar{t} - t' + \hat{t} < \bar{t}$ in contradiction to the minimality of $(\bar{t}, \bar{x})$ for (SP(a,r)). □

Remark 2.22. Thus for a minimal solution $(\bar{t}, \bar{x})$ of the scalar optimization problem (SP(a,r)) with

$$a + \bar{t}r - f(\bar{x}) = \bar{k}, \qquad \bar{k} \neq 0_m,$$

there is a parameter $a' \in H$ and some $t' \in \mathbb{R}$ so that $(t', \bar{x})$ is a minimal solution of (SP(a',r)) with

$$a' + t'r - f(\bar{x}) = 0_m$$

(see Fig. 2.7) and hence $(t', \bar{x})$ is also a minimal solution of $(\overline{\text{SP}}(a', r))$.

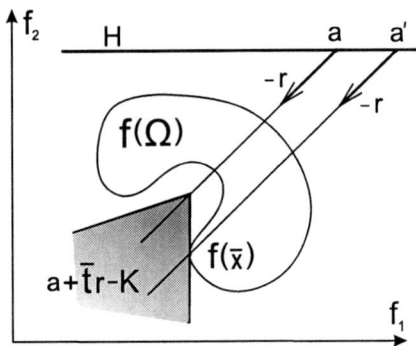

Fig. 2.7. Visualization of the Remark 2.22.

For the scalarization ($\overline{\text{SP}}(a,r)$) the property c) of Theorem 2.1 is no longer valid, which means that minimal solutions of ($\overline{\text{SP}}(a,r)$) are not necessarily weakly K-minimal points of the multiobjective optimization problem (MOP). However due to Theorem 2.21 we can still find all K-minimal points only by varying the parameter a on a hyperplane:

Theorem 2.23. *Let a hyperplane $H = \{y \in \mathbb{R}^m \mid b^\top y = \beta\}$ with $b \in \mathbb{R}^m$, $\beta \in \mathbb{R}$ be given an let $\bar{x} \in \mathcal{M}(f(\Omega), K)$ and $r \in K \setminus \{0_m\}$ with $b^\top r \neq 0$. Then there is a parameter $a \in H$ and some $\bar{t} \in \mathbb{R}$ so that $(\bar{t}, \bar{x})$ is a minimal solution of ($\overline{\text{SP}}(a,r)$).*

Proof. According to Theorem 2.1,b) there is a parameter a' and some $t' \in \mathbb{R}$ so that $(t', \bar{x})$ is a minimal solution of (SP(a', r)). According to Theorem 2.21 there is then a point $a \in H$ and some $\bar{t} \in \mathbb{R}$ so that $(\bar{t}, \bar{x})$ is a minimal solution of ($\overline{\text{SP}}(a,r)$). □

Analogously to the Pascoletti-Serafini method we can do a parameter set restriction for the modified Pascoletti-Serafini problem, too. We demonstrate this for the bicriteria case and we show, that it is again sufficient to consider parameters a from a line segment of the hyperplane H to be able to detect all K-minimal points of the original problem.

Lemma 2.24. *Let $m = 2$ and let the ordering cone K be given as in (2.5). Let $\bar{a}^1$ and $\bar{a}^2$ and the set H^a be given as in Theorem 2.17.*

Then for any K-minimal solution $\bar{x}$ of the multiobjective optimization problem (MOP) there exists a parameter $a \in H^a$ and some $\bar{t} \in \mathbb{R}$ so that $(\bar{t}, \bar{x})$ is a minimal solution of ($\overline{\text{SP}}(a,r)$).

2.4 Modified Pascoletti-Serafini Scalarization

Proof. According to Theorem 2.17 there exists a parameter $a \in H^a$ and some $\bar{t} \in \mathbb{R}$ so that $(\bar{t}, \bar{x})$ is a minimal solution of $(\mathrm{SP}(a,r))$. As shown in the proof of Theorem 2.17 we can choose a and $\bar{t}$ so that $a + \bar{t}r = f(\bar{x})$ and hence, $(\bar{t}, \bar{x})$ is a minimal solution of $(\overline{\mathrm{SP}}(a,r))$, too. □

For the sensitivity studies in the following chapter the Lagrange function and the Lagrange multipliers will be of interest. For that we first recapitulate these notions and then we formulate an extension of Theorem 2.21 taking the Lagrange multipliers into account. We start with a general scalar-valued optimization problem

$$\min F(x)$$
subject to the constraints
$$G(x) \in C,$$
$$H(x) = 0_q,$$
$$x \in S$$

with a closed convex cone $C \subset \mathbb{R}^p$, an open subset $\hat{S} \subset \mathbb{R}^n$, a closed convex set $S \subset \hat{S}$ and continuously differentiable functions $F \colon \hat{S} \to \mathbb{R}$, $G \colon \hat{S} \to \mathbb{R}^p$, and $H \colon \hat{S} \to \mathbb{R}^q$ ($n, p, q \in \mathbb{N}_0$). Then the related Lagrange function is given by $\mathcal{L} \colon \mathbb{R}^n \times C^* \times \mathbb{R}^q \to \mathbb{R}$ (with $C^* = \{y \in \mathbb{R}^p \mid y^\top x \geq 0 \text{ for all } x \in C\}$ the dual cone to C),

$$\mathcal{L}(x, \mu, \xi) := F(x) - \mu^\top G(x) - \xi^\top H(x).$$

If the point x is feasible and if there exists $(\mu, \xi) \in C^* \times \mathbb{R}^q$ with

$$\nabla_x \mathcal{L}(x, \mu, \xi)^\top (s - x) \geq 0 \quad \forall s \in S$$

and

$$\mu^\top G(x) = 0,$$

then μ and ξ are called Lagrange multipliers to the point x.

We need the following assumptions:

Assumption 2.25. Let K be a closed pointed convex cone in $\mathbb{R}^m$ and C a closed convex cone in $\mathbb{R}^p$. Let $\hat{S}$ be a nonempty open subset of $\mathbb{R}^n$ and assume $S \subset \hat{S}$ to be closed and convex. Let the functions $f \colon \hat{S} \to \mathbb{R}^m$, $g \colon \hat{S} \to \mathbb{R}^p$, and $h \colon \hat{S} \to \mathbb{R}^q$ be continuously differentiable on $\hat{S}$.

We now formulate an extended version of Theorem 2.21:

Lemma 2.26. *We consider the scalar optimization problem (SP(a,r)) under the Assumption 2.25. Let $(\bar{t}, \bar{x})$ be a minimal solution and assume there exist Lagrange multipliers $(\mu, \nu, \xi) \in K^* \times C^* \times \mathbb{R}^q$ to the point $(\bar{t}, \bar{x})$. According to Theorem 2.21 there exists a parameter $a' \in H$ and some $t' \in \mathbb{R}$ so that $(t', \bar{x})$ is a minimal solution of (SP(a',r)) and $a' + t'r = f(\bar{x})$.*

Then (μ, ν, ξ) are Lagrange multipliers to the point $(t', \bar{x})$ for the problem (SP(a',r)), too.

Proof. The Lagrange function $\mathcal{L}$ to the scalar optimization problem (SP(a,r)) related to the multiobjective optimization problem (MOP) is given by

$$\mathcal{L}(t, x, \mu, \nu, \xi, a, r) = t - \mu^\top (a + tr - f(x)) - \nu^\top g(x) - \xi^\top h(x).$$

If (μ, ν, ξ) are Lagrange multipliers to the point $(\bar{t}, \bar{x})$ then it follows

$$\nabla_{(t,x)} \mathcal{L}(\bar{t}, \bar{x}, \mu, \nu, \xi, a, r)^\top \begin{pmatrix} t - \bar{t} \\ x - \bar{x} \end{pmatrix} =$$

$$\left[\begin{pmatrix} 1 \\ 0 \end{pmatrix} - \sum_{i=1}^m \mu_i \begin{pmatrix} r_i \\ -\nabla_x f_i(\bar{x}) \end{pmatrix} - \sum_{j=1}^p \nu_j \begin{pmatrix} 0 \\ \nabla_x g_j(\bar{x}) \end{pmatrix} \right.$$
$$\left. - \sum_{k=1}^q \xi_k \begin{pmatrix} 0 \\ \nabla_x h_k(\bar{x}) \end{pmatrix} \right]^\top \begin{pmatrix} t - \bar{t} \\ x - \bar{x} \end{pmatrix} \geq 0 \qquad \forall t \in \mathbb{R}, \ x \in S.$$

Hence $1 - \mu^\top r = 0$ and $\left(\mu^\top \nabla_x f(\bar{x}) - \nu^\top \nabla_x g(\bar{x}) - \xi^\top \nabla_x h(\bar{x})\right)(x - \bar{x}) \geq 0$ for all $x \in S$. Further we have $\mu^\top (a + \bar{t} r - f(\bar{x})) = 0$ and $\nu^\top g(\bar{x}) = 0$. For the minimal solution $(t', \bar{x})$ of the problem (SP(a',r)) it is

$$a' + t' r - f(\bar{x}) = 0_m,$$

and thus $\mu^\top(a' + t'r - f(\bar{x})) = 0$. Because of

$$\nabla_{(t,x)} \mathcal{L}(t', \bar{x}, \mu, \nu, \xi, a', r) = \nabla_{(t,x)} \mathcal{L}(\bar{t}, \bar{x}, \mu, \nu, \xi, a, r)$$

we also have

$$\nabla_{(t,x)} \mathcal{L}(t', \bar{x}, \mu, \nu, \xi, a', r)^\top \begin{pmatrix} t - t' \\ x - \bar{x} \end{pmatrix} \geq 0 \qquad \forall t \in \mathbb{R}, \ x \in S.$$

Thus (μ, ν, ξ) are also Lagrange multipliers to the point $(t', \bar{x})$ for the problem (SP(a',r)). □

In the chapter about sensitivity considerations the Lagrange multipliers play an important role.

2.5 Relations Between Scalarizations

We have seen in the preceding sections that the Pascoletti-Serafini problem features many interesting and important properties. It is a very general formulation allowing two parameters to vary arbitrarily. Due to this, many other well-known and wide spread scalarization approaches can be seen as a special case and can be subsumed under this general problem. The connections will be studied in this section. The relations are important for applying the results about an adaptive parameter control gained in the following chapters for the general scalarization to the special problems, too.

2.5.1 ε-Constraint Problem

We start with a common method called ε-constraint method ([54, 98, 60, 159, 165]). It is a very wide spread method especially in engineering design for finding EP-minimal points, because the method is very intuitive and the parameters are easy to interpret as upper bounds. In [147, 148, 186] this method is used for solving multiobjective optimization problems via evolutionary algorithms.

For an arbitrary $k \in \{1, \ldots, m\}$ and parameters $\varepsilon_i \in \mathbb{R}$, $i \in \{1, \ldots, m\} \setminus \{k\}$, the scalarized problem called $(P_k(\varepsilon))$ reads as follows (compare (2.1)):

$$
\begin{aligned}
&\min f_k(x) \\
&\text{subject to the constraints} \\
&f_i(x) \leq \varepsilon_i, \quad i \in \{1, \ldots, m\} \setminus \{k\}, \\
&x \in \Omega.
\end{aligned}
\quad (2.24)
$$

It is easy to see that this is just a special case of the Pascoletti-Serafini scalarization for the ordering cone $K = \mathbb{R}^m_+$. We even get a connection w.r.t. the Lagrange multipliers:

Theorem 2.27. *Let Assumption 2.25 hold and let $K = \mathbb{R}^m_+$, $C = \mathbb{R}^p_+$, and $\hat{S} = S = \mathbb{R}^n$. A point $\bar{x}$ is a minimal solution of $(P_k(\varepsilon))$ with Lagrange multipliers $\bar{\mu}_i \in \mathbb{R}_+$ for $i \in \{1, \ldots, m\} \setminus \{k\}$, $\bar{\nu} \in \mathbb{R}^p_+$, and $\bar{\xi} \in \mathbb{R}^q$, if and only if $(f_k(\bar{x}), \bar{x})$ is a minimal solution of $(SP(a,r))$ with Lagrange multipliers $(\bar{\mu}, \bar{\nu}, \bar{\xi})$ with $\bar{\mu}_k = 1$, and*

$$a_i = \varepsilon_i, \quad \forall i \in \{1, \ldots, m\} \setminus \{k\}, \quad a_k = 0 \text{ and } r = e_k \quad (2.25)$$

with e_k the kth unit vector in $\mathbb{R}^m$.

Proof. By introducing the additional variable $t \in \mathbb{R}$ the scalar optimization problem $(P_k(\varepsilon))$ can be formulated as

$$\min t$$

subject to the constraints
$$\begin{aligned}
\varepsilon_i - f_i(x) &\geq 0, & i &\in \{1, \ldots, m\} \setminus \{k\}, \\
t - f_k(x) &\geq 0, & & \quad\quad\quad\quad\quad\quad (2.26)\\
g_j(x) &\geq 0, & j &= 1, \ldots, p, \\
h_l(x) &= 0, & l &= 1, \ldots, q, \\
t \in \mathbb{R}, \; x &\in \mathbb{R}^n.
\end{aligned}$$

If $\bar{x}$ is a minimal solution of $(P_k(\varepsilon))$ then $(\bar{t}, \bar{x}) := (f_k(\bar{x}), \bar{x})$ is a minimal solution of the problem (2.26). However problem (2.26) is equivalent to the Pascoletti-Serafini problem $(SP(a,r))$ with a and r as in (2.25).

Because $\bar{\mu}_i$, $i \in \{1, \ldots, m\} \setminus \{k\}$, $\bar{\nu}_j$, $j = 1, \ldots, p$, $\bar{\xi}_l$, $l = 1, \ldots, q$, are Lagrange multipliers to $\bar{x}$ for $(P_k(\varepsilon))$, we have

$$\begin{aligned}
\bar{\mu}_i(\varepsilon_i - f_i(\bar{x})) &= 0 & &\text{for all } i \in \{1, \ldots, m\} \setminus \{k\}, \\
\bar{\nu}_j(g_j(\bar{x})) &= 0 & &\text{for all } j \in \{1, \ldots, p\},
\end{aligned}$$

and

$$\nabla f_k(\bar{x}) + \sum_{\substack{i=1 \\ i \neq k}}^m \bar{\mu}_i \nabla f_i(\bar{x}) - \sum_{j=1}^p \bar{\nu}_j \nabla g_j(\bar{x}) - \sum_{l=1}^q \bar{\xi}_l \nabla h_l(\bar{x}) = 0_n. \quad (2.27)$$

The derivative of the Lagrange function $\mathcal{L}(t, x, \mu, \nu, \xi, a, r)$ to $(SP(a,r))$ with a and r as in (2.25) in the point $(f_k(\bar{x}), \bar{x})$ reads as follows:

$$\nabla_{(t,x)} \mathcal{L}(f_k(\bar{x}), \bar{x}, \mu, \nu, \xi, a, r) = \begin{pmatrix} 1 \\ 0 \end{pmatrix} - \mu_k \begin{pmatrix} 1 \\ -\nabla f_k(\bar{x}) \end{pmatrix}$$
$$- \sum_{\substack{i=1 \\ i \neq k}}^m \mu_i \begin{pmatrix} 0 \\ -\nabla f_i(\bar{x}) \end{pmatrix} - \sum_{j=1}^p \nu_j \begin{pmatrix} 0 \\ \nabla g_j(\bar{x}) \end{pmatrix} - \sum_{l=1}^q \xi_l \begin{pmatrix} 0 \\ \nabla h_l(\bar{x}) \end{pmatrix}.$$

By choosing $\bar{\mu}_k = 1$ and applying (2.27) we get

$$\nabla_{(t,x)} \mathcal{L}(f_k(\bar{x}), \bar{x}, \bar{\mu}, \bar{\nu}, \bar{\xi}, a, r) = 0_{n+1},$$

and hence $(\bar{\mu}, \bar{\nu}, \bar{\xi})$ are Lagrange multipliers to the point $(f_k(\bar{x}), \bar{x})$ for the problem $(SP(a,r))$, too. The proof of the converse direction can be done analogously. $\square$

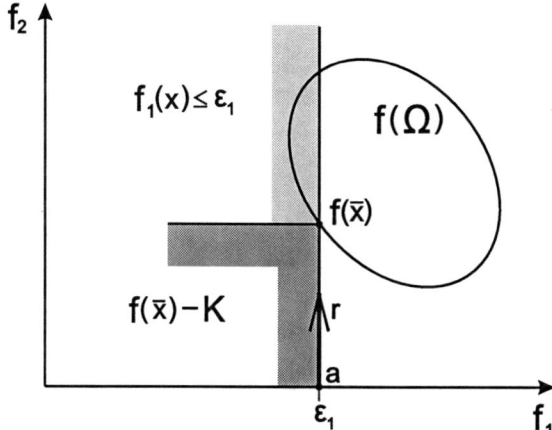

Fig. 2.8. Connection between the ε-constraint and the Pascoletti-Serafini method.

The statement of Theorem 2.27 is visualized in Fig. 2.8 on a bicriteria optimization problem with $k = 2$ in the ε-constraint method.

For the choice of the parameters as in (2.25) it follows that the constraint $a_k + t r_k - f_k(x) \geq 0$ is always active in $(f_k(\bar{x}), \bar{x})$, i.e. it is $a_k + t r_k - f_k(x) = 0$. The ε-constraint method is a restriction of the Pascoletti-Serafini problem with the parameter a chosen only from the hyperplane $H = \{y \in \mathbb{R}^m \mid y_k = 0\}$ and the parameter $r = e_k$ constant. From Theorem 2.27 together with Theorem 2.11 we conclude:

Corollary 2.28. *If $\bar{x} \in \mathcal{M}(f(\Omega), \mathbb{R}^m_+)$, then $\bar{x}$ is a minimal solution of $(P_k(\varepsilon))$ for $\varepsilon_i = f_i(\bar{x})$, $i \in \{1, \ldots, m\} \setminus \{k\}$.*

In contrast to the Pascoletti-Serafini method in general not any weakly EP-minimal solution can be found by solving $(P_k(\varepsilon))$ because we choose $r \in \partial K = \partial \mathbb{R}^m_+$ for the ε-constraint method. However weakly EP-minimal points which are not also EP-minimal are not of practical interest. As a consequence of Theorem 2.1,c) we have:

Corollary 2.29. *If $\bar{x}$ is a minimal solution of $(P_k(\varepsilon))$, then $\bar{x} \in \mathcal{M}_w(f(\Omega), \mathbb{R}^m_+)$.*

A direct proof of this result can be found for instance in [62, Prop. 4.3] or in [165, Theorem 3.2.1].

The ε-constraint method has a big drawback against the more general Pascoletti-Serafini problem. According to Corollary 2.5, if we have

$\mathcal{E}(f(\Omega), \mathbb{R}_+^m) \neq \emptyset$ and $f(\Omega)+K$ is closed and convex, then for any choice of the parameters $(a, r) \in \mathbb{R}^m \times \text{int}(K)$ there exists a minimal solution of the problem (SP(a,r)). This is no longer true for the ε-constraint method as the following example demonstrates.

Example 2.30. Let $f \colon \mathbb{R}^2 \to \mathbb{R}^2$ with $f(x) := x$ for all $x \in \mathbb{R}^2$ be given. We consider the bicriteria optimization problem

$$\min f(x) = x$$
$$\text{subject to the constraints}$$
$$\|x\|_2 \leq 1,$$
$$x \in \mathbb{R}^2$$

w.r.t. the natural ordering $K = \mathbb{R}_+^m$. The set $f(\Omega) + K$ is convex and the efficient set is

$$\mathcal{E}(f(\Omega), \mathbb{R}_+^2) = \{x = (x_1, x_2)^\top \in \mathbb{R}^2 \mid \|x\|_2 = 1,\ x_1 \leq 0,\ x_2 \leq 0\} \neq \emptyset.$$

The ε-constraint scalarization for $k = 2$ is given by

$$\min f_2(x)$$
$$\text{subject to the constraints}$$
$$f_1(x) \leq \varepsilon_1,$$
$$\|x\|_2 \leq 1,$$
$$x \in \mathbb{R}^2,$$

but for $\varepsilon_1 < -1$ there exists no feasible point and thus no minimal solution.

Hence it can happen that the ε-constraint problem is solved for a large number of parameters without getting any solution, and with that weakly EP-minimal points, or at least the information $\mathcal{M}(f(\Omega), \mathbb{R}_+^m) = \emptyset$. This is due to the fact that this is a special case of the Pascoletti-Serafini problem with $r \in \partial K = \partial \mathbb{R}_+^m$.

We can apply the results of Theorem 2.17 for a restriction of the parameter set for the ε-constraint problem, too. For $m = 2$ and e.g. $k = 2$ we have according to Theorem 2.27

$$r = \begin{pmatrix} 0 \\ 1 \end{pmatrix} \quad \text{and} \quad a \in H = \{y \in \mathbb{R}^2 \mid y_2 = 0\} = \left\{ \begin{pmatrix} \varepsilon \\ 0 \end{pmatrix} \,\middle|\, \varepsilon \in \mathbb{R} \right\}$$

(i.e. $b = (0,1)^\top$, $\beta = 0$). The ordering cone $K = \mathbb{R}_+^2$ is finitely generated by $l^1 = (1,0)^\top$ and $l^2 = (0,1)^\top$ and hence (2.6) and (2.7) are equal to

$$\min_{x \in \Omega} f_1(x) \quad \text{and} \quad \min_{x \in \Omega} f_2(x).$$

For $\bar{x}^1$ and $\bar{x}^2$ respectively minimal solutions of these problems we get

$$H^a := \{y \in H \mid y = \lambda \bar{a}^1 + (1-\lambda)\bar{a}^2, \ \lambda \in [0,1]\}$$

with

$$\bar{a}^i := f(\bar{x}^i) - \frac{b^\top f(\bar{x}^i) - \beta}{b^\top r} r = \begin{pmatrix} f_1(\bar{x}^i) \\ 0 \end{pmatrix}, \ i = 1, 2,$$

and hence

$$\begin{aligned} H^a &= \{y = (\varepsilon, 0)^\top \mid \varepsilon = \lambda f_1(\bar{x}^1) + (1-\lambda) f_1(\bar{x}^2), \ \lambda \in [0,1]\} \\ &= \{y = (\varepsilon, 0)^\top \mid f_1(\bar{x}^1) \le \varepsilon \le f_1(\bar{x}^2)\}. \end{aligned}$$

We conclude:

Corollary 2.31. *Let $\bar{x}$ be an EP-minimal solution of the multiobjective optimization problem (MOP) with $m = 2$. Let $\bar{x}^1$ be a minimal solution of $\min_{x \in \Omega} f_1(x)$ and $\bar{x}^2$ a minimal solution of $\min_{x \in \Omega} f_2(x)$. Then there is a parameter $\varepsilon \in \{y \in \mathbb{R} \mid f_1(\bar{x}^1) \le y \le f_1(\bar{x}^2)\}$ with $\bar{x}$ a minimal solution of $(P_2(\varepsilon))$. The same result holds for $(P_1(\varepsilon))$.*

2.5.2 Normal Boundary Intersection Problem

We start with a short recapitulation of this method introduced by Das and Dennis in [38, 40]. For determining EP-minimal points the scalar optimization problems

$$\begin{aligned} &\max s \\ &\text{subject to the constraints} \\ &\Phi \beta + s \bar{n} = f(x) - f^*, \\ &s \in \mathbb{R}, \ x \in \Omega, \end{aligned} \qquad (2.28)$$

named (NBI(β)) for parameters $\beta \in \mathbb{R}_+^m$, $\sum_{i=1}^m \beta_i = 1$, are solved. Here f^* denotes the so-called ideal point defined by $f_i^* := f_i(x^i) := \min_{x \in \Omega} f_i(x)$, $i = 1, \ldots, m$. The matrix $\Phi \in \mathbb{R}^{m \times m}$ consists of the columns $f(x^i) - f^*$ $(i = 1, \ldots, m)$ and the set

$$f^* + \{\Phi\beta \mid \beta \in \mathbb{R}_+^m, \sum_{i=1}^m \beta_i = 1\} \qquad (2.29)$$

is then the set of all convex combinations of the extremal points $f(x^i)$, $i = 1, \ldots, m$, the so-called CHIM (convex hull of individual minima). The vector $\bar{n}$ is defined as normal unit vector to the hyperplane extending the CHIM directing to the negative orthant.

The idea of this method is that by solving the problem (NBI(β)) for an equidistant choice of parameters β an equidistant approximation of the efficient set is generated. However already for the case $m \geq 3$ generally not all EP-minimal points can be found as a solution of (NBI(β)) (see [40, Fig. 3]) and what is more, the maximal solutions of (NBI(β)) are not necessarily weakly EP-minimal ([231, Ex. 7.9.1]).

There is a direct connection between the normal boundary intersection (NBI) method and the modified version $(\overline{SP}(a,r))$ of the Pascoletti-Serafini problem.

Lemma 2.32. *A point $(\bar{s}, \bar{x})$ is a maximal solution of (NBI(β)) with $\beta \in \mathbb{R}^m$, $\sum_{i=1}^m \beta_i = 1$, if and only if $(-\bar{s}, \bar{x})$ is a minimal solution of $(\overline{SP}(a,r))$ with $a = f^* + \Phi\beta$ and $r = -\bar{n}$.*

Proof. By setting $a = f^* + \Phi\beta$, $t = -s$ and $r = -\bar{n}$ we see immediately that solving problem (NBI(β)) is equivalent to solve

$$-\min t$$

subject to the constraints

$$a + tr - f(x) = 0_m,$$
$$t \in \mathbb{R}, \ x \in \Omega,$$

being again equivalent to solve $(\overline{SP}(a,r))$. □

Hence, the NBI method is a restriction of the modified Pascoletti-Serafini method as the parameter a is chosen only from the CHIM and the parameter $r = -\bar{n}$ is chosen constant. In the bicriteria case ($m = 2$) the CHIM consists of the points:

$$f^* + \Phi\beta = \begin{pmatrix} f_1(x^1) \\ f_2(x^2) \end{pmatrix} + \begin{pmatrix} 0 & f_1(x^2) - f_1(x^1) \\ f_2(x^1) - f_2(x^2) & 0 \end{pmatrix} \begin{pmatrix} \beta_1 \\ 1 - \beta_1 \end{pmatrix}$$
$$= \beta_1 \begin{pmatrix} f_1(x^1) \\ f_2(x^1) \end{pmatrix} + (1 - \beta_1) \begin{pmatrix} f_1(x^2) \\ f_2(x^2) \end{pmatrix}$$

2.5 Relations Between Scalarizations

for $\beta = (\beta_1, 1 - \beta_1)^\top \in \mathbb{R}^2_+$. Then the hyperplane H including the CHIM is

$$H := \{f^* + \Phi\beta \mid \beta \in \mathbb{R}^2, \ \beta_1 + \beta_2 = 1\}$$
$$= \{\beta f(x^1) + (1 - \beta) f(x^2) \mid \beta \in \mathbb{R}\}.$$

The set H^a according to Lemma 2.24 is then

$$H^a = \{\beta f(x^1) + (1 - \beta) f(x^2) \mid \beta \in [0, 1]\}$$

which equals (only here in the bicriteria case) the set CHIM proposed by Das and Dennis.

In the case of more then three objective functions it is no longer sufficient to consider convex combinations of the extremal points as we have already seen in Example 2.19. That is the reason why in general not all EP-minimal points can be found with the NBI method (as proposed by Das and Dennis) for the case $m \geq 3$. However by allowing the parameter β to vary arbitrarily, and with that the parameter a to vary arbitrarily on the hyperplane including the CHIM, all EP-minimal points of (MOP) can be found by solving ($\overline{\text{SP}}(a, r)$) and (NBI($\beta$)) respectively (see Theorem 2.23).

For a discussion of the NBI method see also [140, 209]. In [208] a modification of the NBI method is proposed. There, the equality constraint in (NBI(β)) is replaced by the inequality

$$\Phi\beta + s\bar{n} \geq_m f(x) - f^*.$$

This modified problem guarantees thus weakly efficient points. It is then a special case of the Pascoletti-Serafini scalarization (SP(a, r)) with the parameters as in Lemma 2.32. In [208] also the connection between that modified problem and the weighted sum as well as the ε-constraint problem are discussed.

2.5.3 Modified Polak Problem

The modified Polak method ([112, 128, 182] and an application in [127]) has a similar connection to the Pascoletti-Serafini problem as the normal boundary intersection method has. We restrict the presentation of the modified Polak method here to the bicriteria case. Then, for different values of the parameter $y_1 \in \mathbb{R}$, the scalar optimization problems called (MP(y_1))

$$\min f_2(x)$$
$$\text{subject to the constraints}$$
$$f_1(x) = y_1,$$
$$x \in \Omega$$
(2.30)

are solved. Here the objectives are transformed to equality constraints, like in the normal boundary intersection problem, and not to inequality constraints, like in the general Pascoletti-Serafini method. Besides the constraint $f_1(x) = y_1$ shows a similarity to the ε-constraint method with the constraint $f_1(x) \leq \varepsilon_1$.

Lemma 2.33. *Let $m = 2$. A point $\bar{x}$ is a minimal solution of $(MP(y_1))$ if and only if $(f_2(\bar{x}), \bar{x})$ is a minimal solution of $(\overline{SP}(a,r))$ with $a = (y_1, 0)$ and $r = (0, 1)$.*

Proof. With the parameters a and r as defined in the theorem problem $(\overline{SP}(a,r))$ reads as follows:

$$\min t$$
$$\text{subject to the constraints}$$
$$y_1 - f_1(x) = 0,$$
$$t - f_2(x) = 0,$$
$$t \in \mathbb{R}, \ x \in \Omega,$$

and it can immediately be seen that solving this problem is equivalent to solve problem $(MP(y_1))$. □

Of course a generalization to the case $m \geq 3$ can be done as well.

The modified Polak method is, like the NBI method, a restriction of the modified Pascoletti-Serafini method. However because the parameter $a = (y_1, 0)$ is allowed to vary arbitrarily in the hyperplane $H = \{y \in \mathbb{R}^2 \mid y_2 = 0\}$ in contrast to the NBI method all EP-minimal points can be found. We can apply Lemma 2.24 and get again a result on the restriction of the parameter set:

Lemma 2.34. *Let $m = 2$, $K = \mathbb{R}^2_+$ and let $f_1(\bar{x}^1) := \min_{x \in \Omega} f_1(x)$ and $f_2(\bar{x}^2) := \min_{x \in \Omega} f_2(x)$ be given. Then, for any EP-minimal solution $\bar{x}$ of the multiobjective optimization problem (MOP) there exists a parameter y_1 with $f_1(\bar{x}^1) \leq y_1 \leq f_1(\bar{x}^2)$ such that $\bar{x}$ is a minimal solution of $(MP(y_1))$.*

Proof. We determine the set H^a according to Lemma 2.24. It is $H = \{y \in \mathbb{R}^2 \mid (0,1)y = 0\}$ and $r = (0,1)^\top$. Then it follows

$$\bar{a}^1 = \begin{pmatrix} f_1(\bar{x}^1) \\ 0 \end{pmatrix} \quad \text{and} \quad \bar{a}^2 = \begin{pmatrix} f_1(\bar{x}^2) \\ 0 \end{pmatrix}$$

and thus $H^a = \{y = (y_1, 0) \in \mathbb{R}^2 \mid f_1(\bar{x}^1) \leq y_1 \leq f_2(\bar{x}^2)\}$. As a consequence of Lemma 2.24 there exists some $\bar{t} \in \mathbb{R}$ and a parameter $a = (y_1, 0) \in H^a$ with $(\bar{t}, \bar{x})$ a minimal solution of $(\overline{SP}(a,r))$. Here, $\bar{t}$ is according to the proof of Theorem 2.17 given as $\bar{t} = f_2(\bar{x})$. Thus $(f_2(\bar{x}), \bar{x})$ is a minimal solution of $(\overline{SP}(a,r))$ and with Lemma 2.33 we conclude that $\bar{x}$ is a minimal solution of $(MP(y_1))$ with $f_1(\bar{x}^1) \leq y_1 \leq f_1(\bar{x}^2)$. □

This result is also used in the algorithm for the modified Polak method presented in [124, p.314].

2.5.4 Weighted Chebyshev Norm Problem

In this scalarization method ([56, 151, 158, 216], [112, p.13]) for determining EP-minimal points we have weights $w_i > 0$, $i = 1, \ldots, m$, and a reference point ([25, 167, 236]) $a \in \mathbb{R}^m$ with $a_i < \min_{x \in \Omega} f_i(x)$, $i = 1, \ldots, m$, (assuming solutions exist), i.e. $f(\Omega) \subset a + \text{int}(\mathbb{R}^m_+)$, as parameters. For scalarizing the multiobjective optimization problem the weighted Chebyshev norm of the function $f(\cdot) - a$ is minimized:

$$\min_{x \in \Omega} \max_{i \in \{1,\ldots,m\}} w_i(f_i(x) - a_i). \tag{2.31}$$

This problem has the following connection to the Pascoletti-Serafini problem:

Theorem 2.35. *A point $(\bar{t}, \bar{x}) \in \mathbb{R} \times \Omega$ is a minimal solution of $(SP(a,r))$ with $K = \mathbb{R}^m_+$ and with parameters $a \in \mathbb{R}^m$, $a_i < \min_{x \in \Omega} f_i(x)$, $i = 1, \ldots, m$, and $r \in \text{int}(\mathbb{R}^m_+)$ if and only if $\bar{x}$ is a solution of (2.31) with reference point a and weights $w_i = \frac{1}{r_i} > 0$, $i = 1, \ldots, m$.*

Proof. If we set $r_i = \frac{1}{w_i} > 0$ and $K = \mathbb{R}^m_+$ problem $(SP(a,r))$ reads as follows:

$$\min t$$
subject to the constraints
$$a_i + t \tfrac{1}{w_i} - f_i(x) \geq 0, \quad i = 1, \ldots, m,$$
$$t \in \mathbb{R}, \; x \in \Omega.$$

This is because of $w_i > 0$, $i = 1, \ldots, m$, equivalent to

$$\min t$$
subject to the constraints
$$w_i(f_i(x) - a_i) \leq t, \quad i = 1, \ldots, m,$$
$$t \in \mathbb{R}, \ x \in \Omega,$$

but this is according to [124, p.305], or [151, p.14] just a reformulation of (2.31) with an additional variable introduced. □

Thus a variation of the weights in the norm corresponds to a variation of the direction r, and a variation of the reference point is like a variation of the parameter a in the Pascoletti-Serafini method. Similar in the goal attainment method by Gembicki and Haimes, [89], for determining EP-minimal points for the case $m = 2$ the parameter a is interpreted as a goal and the parameter r with $r_i > 0$, $i = 1, 2$, as weights of the deviation of the objective function to the goal. For generating various efficient points the parameter $r \in \text{int}(\mathbb{R}_+^m)$ with $r_1 + r_2 = 1$ is varied.

As a consequence of Theorem 2.1 any solution of (2.31) is at least weakly EP-minimal. Besides, any weakly EP-minimal point can be found as a solution of (2.31). In [231, Example 7.7.1] it is shown, that $f(\Omega) \subset a + \mathbb{R}_+^m$ is necessary for the last statement.

For $a = 0_m$ problem (2.31) reduces to the weighted minimax method as discussed in [149].

2.5.5 Problem According to Gourion and Luc

This problem is lately developed by Gourion and Luc, [95], for finding EP-maximal points of a multiobjective optimization problem with $f(\Omega) \subset \mathbb{R}_+^m$. This corresponds to the multiobjective optimization problem (MOP) w.r.t. the ordering cone $K = \mathbb{R}_-^m$. The parameter dependent scalar optimization problems according to Gourion and Luc read as follows

$$\max s$$
subject to the constraints
$$f(x) \geq s\alpha,$$
$$s \in \mathbb{R}, \ x \in \Omega$$
(2.32)

introducing the new variable $s \in \mathbb{R}$ and with the parameter $\alpha \in \mathbb{R}_+^m$. We will see that the scalarization of Gourion and Luc can be seen as a special case of the Pascoletti-Serafini method with a variation of the parameter $r = -\alpha$ only and with the constant parameter $a = 0_m$.

Theorem 2.36. *A point $(\bar{s}, \bar{x})$ is a maximal solution of (2.32) with parameter $\alpha \in \mathbb{R}_+^m$ if and only if $(-\bar{s}, \bar{x})$ is a minimal solution of $(SP(a, r))$ with $a = 0_m$, $r = -\alpha \in \mathbb{R}_-^m$ and $K = \mathbb{R}_-^m$.*

Proof. By defining $r = -\alpha \in \mathbb{R}_-^m$ and $t = -s$ problem (2.32) can be written as

$$\max (-t)$$
subject to the constraints
$$f(x) \geq (-r) \cdot (-t),$$
$$t \in \mathbb{R}, \ x \in \Omega$$

being equivalent to

$$- \min t$$
subject to the constraints
$$tr - f(x) \in K,$$
$$t \in \mathbb{R}, \ x \in \Omega$$

with $K = \mathbb{R}_-^m$, i.e. to the Pascoletti-Serafini scalarization $(SP(a, r))$ with $a = 0_m$ and $K = \mathbb{R}_-^m$. □

From this theorem together with Theorem 2.1,c) it follows that if $(\bar{s}, \bar{x})$ is a maximal solution of (2.32) then $\bar{x}$ is weakly $\mathbb{R}_-^m$-minimal, i.e. $\bar{x}$ is weakly EP-maximal for the related multiobjective optimization problem.

For the choice of the parameter $\alpha = -r$ Gourion and Luc present a procedure and give a convergence proof for this method. Further for special sets $f(\Omega)$ they show that the minimal value function (in the notion of the Pascoletti-Serafini problem)

$$r \mapsto \min\{t \mid tr - f(x) \in K, \ t \in \mathbb{R}, \ x \in \Omega\}$$

is continuous on the set $\{r \in \mathbb{R}_+^m \mid \sum_{i=1}^{m} r_i = 1\}$ ([95, Lemma 3.1]).

2.5.6 Generalized Weighted Sum Problem

Before we come to the usual weighted sum method we consider a more general formulation having not only a weighted sum as objective function but also similar constraints as the already discussed ε-constraint method (compare [231, p.136]):

$$\min \sum_{i=1}^{m} w_i f_i(x) = w^\top f(x)$$
subject to the constraints (2.33)
$$f_i(x) \leq \varepsilon_i \quad \text{for all } i \in P,$$
$$x \in \Omega$$

with $P \subsetneq \{1,\ldots,m\}$, $\varepsilon_i \in \mathbb{R}$ for all $i \in P$, and weights $w \in \mathbb{R}^m$, $\sum_{i \notin P} w_i > 0$. We start with the connection to the Pascoletti-Serafini problem and from that we conclude some properties of the problem (2.33).

Theorem 2.37. *A point $\bar{x}$ is a minimal solution of (2.33) for the parameters $w \in \mathbb{R}^m$, $\sum_{i \notin P} w_i > 0$, $\varepsilon_i \in \mathbb{R}$, $i \in P$, if and only if there is some $\bar{t}$ so that $(\bar{t}, \bar{x})$ is a minimal solution of $(SP(a,r))$ with $a_i = \varepsilon_i$ for $i \in P$, a_i arbitrary for $i \in \{1,\ldots,m\} \setminus P$, $r_i = 0$ for $i \in P$, $r_i = 1$ for $i \in \{1,\ldots,m\} \setminus P$ and cone $K_w := \{y \in \mathbb{R}^m \mid y_i \geq 0, \text{ for all } i \in P, w^\top y \geq 0\}$, i. e. of*

$$\min t$$
subject to the constraints (2.34)
$$a + tr - f(x) \in K_w,$$
$$t \in \mathbb{R}, \; x \in \Omega.$$

Proof. The optimization problem (2.34) is equivalent to

$$\min t$$
subject to the constraints
$$w^\top (a + tr - f(x)) \geq 0, \quad (2.35)$$
$$a_i + tr_i - f_i(x) \geq 0 \quad \text{for all } i \in P,$$
$$t \in \mathbb{R}, \; x \in \Omega.$$

As $a_i = \varepsilon_i$ and $r_i = 0$ for $i \in P$ and because of $w^\top r = \sum_{i \notin P} w_i > 0$, a point $(\bar{t}, \bar{x})$ is a minimal solution of (2.35) if and only if $\bar{x}$ is a minimal solution of

$$\min \frac{w^\top f(x) - w^\top a}{w^\top r}$$
subject to the constraints (2.36)
$$f_i(x) \leq \varepsilon_i \quad \text{for all } i \in P,$$
$$x \in \Omega.$$

Because we can ignore the constant term $-\frac{w^T a}{w^T r}$ in the objective function of (2.36) and because of $w^T r > 0$ a point $\bar{x}$ is a minimal solution of (2.36) if and only if it is a minimal solution of (2.33). □

The set K_w is a closed convex cone and for $|P| = m - 1$ the cone K_w is even pointed ([231, Lemma 7.11.1]).

Corollary 2.38. *Let $\bar{x}$ be a minimal solution of (2.33) and let $K \subset \{y \in \mathbb{R}^m \mid y_i \geq 0 \text{ for all } i \in P\}$, $w \in K^* \setminus \{0_m\}$ and $\sum_{i \notin P} w_i > 0$, then $\bar{x} \in \mathcal{M}_w(f(\Omega), K)$.*

Proof. Applying Theorem 2.37 and Theorem 2.1,c) it follows that $\bar{x} \in \mathcal{M}_w(f(\Omega), K_w)$ with $K_w = \{y \in \mathbb{R}^m \mid y_i \geq 0 \text{ for all } i \in P, w^T y \geq 0\}$. As for all $y \in K$ we have $w^T y \geq 0$ it follows $K \subset K_w$ and hence, with Lemma 1.7, $\mathcal{M}_w(f(\Omega), K_w) \subset \mathcal{M}_w(f(\Omega), K)$. □

Thus, e. g. for $K = \mathbb{R}^m_+$, $w \in \mathbb{R}^m_+$ and $w_i > 0$ for all $i \notin P$, all minimal solutions of (2.33) are at least weakly EP-minimal. For $w_i > 0$, $i = 1, \ldots, m$, they are even EP-minimal (see [231, Theorem 7.11.1c)]).

Of course we can find all EP-minimal points $\bar{x} \in \mathcal{M}(f(\Omega), \mathbb{R}^m_+)$ by solving (2.33) if we set $P = \{1, \ldots, m-1\}$, $w_i = 0$ for $i \in P$, and $w_m = 1$. Then $K_w = \mathbb{R}^m_+$ and (2.33) equals the ε-constraint problem $(P_m(\varepsilon))$. By choosing $\varepsilon_i = f_i(\bar{x})$ for all $i \in P$, it is known that $\bar{x}$ is a minimal solution of the ε-constraint problem and hence of the problem (2.33) (see Corollary 2.31).

The case $P = \{1, \ldots, m\}$ is introduced and discussed in Charnes and Cooper ([29]) and later in Wendell and Lee ([233]). See also [97]. Weidner has shown ([231, p.130]) that there is no equivalent formulation between (2.33) with $P = \{1, \ldots, m\}$ and $(SP(a, r))$.

2.5.7 Weighted Sum Problem

Now we come to the usual weighted sum method ([245], see also [55, 86])

$$\min w^T f(x)$$
$$\text{subject to the constraint} \qquad (2.37)$$
$$x \in \Omega$$

for weights $w \in K^* \setminus \{0_m\}$ which is just a special case of (2.33) for $P = \emptyset$. Because it is such an important problem formulation we adapt Theorem 2.37 for this special case (see also Fig. 2.9):

Theorem 2.39. *A point $\bar{x}$ is a minimal solution of (2.37) for the parameter $w \in K^* \setminus \{0_m\}$ if and only if there is some $\bar{t}$ so that $(\bar{t}, \bar{x})$ is a minimal solution of (SP(a,r)) with $a \in \mathbb{R}^m$ arbitrarily chosen, cone $K_w := \{y \in \mathbb{R}^m \mid w^\top y \geq 0\}$ and $r \in \text{int}(K_w)$.*

Proof. Problem (SP(a,r)) with cone K_w reads as follows:

$$\begin{aligned} &\min t \\ &\text{subject to the constraints} \\ &w^\top(a + tr - f(x)) \geq 0, \\ &t \in \mathbb{R}, \ x \in \Omega. \end{aligned} \qquad (2.38)$$

Because of $r \in \text{int}(K_w)$ we have $w^\top r > 0$ and hence (2.38) is equivalent to

$$\min \frac{w^\top f(x) - w^\top a}{w^\top r}$$

subject to the constraint

$$x \in \Omega.$$

With the same arguments as used in the proof to Theorem 2.37 this is equivalent to (2.37). □

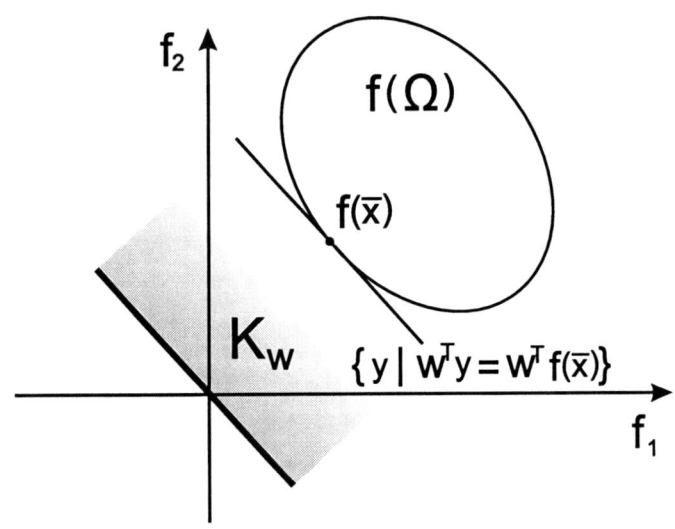

Fig. 2.9. Connection between the weighted sum and the Pascoletti-Serafini problem.

2.5 Relations Between Scalarizations   63

Hence a variation of the weights $w \in K^* \setminus \{0_m\}$ corresponds to a variation of the ordering cone K_w. So we get a new interpretation for the weighted sum method. The cone K_w is a closed convex polyhedral cone, but K_w is not pointed. That is the reason why the results from Theorem 2.1,b) cannot be applied to the weighted sum method and why it is in general (in the non-convex case) not possible to find all K-minimal points of (MOP) by solving the weighted sum method with appropriate weights. However it is known that in the case of a convex set $f(\Omega)$ all K-minimal points of the multiobjective optimization problem (MOP) can be found ([124, pp.299f]). In [239] the stability of the solutions of the weighted sum problem is studied.

We can conclude from Theorem 2.1,c) the following well known result:

Corollary 2.40. *Let $\bar{x}$ be a minimal solution of (2.37) with parameter $w \in K^* \setminus \{0_m\}$, then $\bar{x}$ is weakly K-minimal for (MOP).*

Proof.   According to Theorem 2.39 there is some $\bar{t}$ so that $(\bar{t}, \bar{x})$ is a minimal solution of $(\mathrm{SP}(a,r))$ with cone K_w and hence $\bar{x}$ is according to Theorem 2.1,c) and Remark 2.2 a weakly K_w-minimal point. Because $w \in K^* \setminus \{0_m\}$ we have $w^\top y \geq 0$ for all $y \in K$ and hence $K \subset K_w$. Thus, according to Lemma 1.7, $\mathcal{M}_w(f(\Omega), K_w) \subset \mathcal{M}_w(f(\Omega), K)$ and hence $\bar{x}$ is a weakly K-minimal point, too. □

The weighted sum method has the same drawback against the Pascoletti-Serafini method as the ε-constraint method has, as shown in Example 2.30: not for any choice of the parameters $w \in K^* \setminus \{0_m\}$ there exists a minimal solution, even not for the case $\mathcal{M}(f(\Omega), K) \neq \emptyset$. In [24, Ex. 7.3] Brosowski gives a simple example where the weighted sum problem delivers only for one choice of weights a minimal solution and where it is not solvable for all other weights despite the set $f(\Omega) + K$ is closed and convex in contrast to Corollary 2.5:

Example 2.41. We consider the bicriteria optimization problem

$$\min f(x) = x$$
$$\text{subject to the constraints}$$
$$x_1 + x_2 \geq 1,$$
$$x \in \mathbb{R}^2$$

w. r. t. the natural ordering. The set of K-minimal points is given by

$$\{x = (x_1, x_2)^\top \in \mathbb{R}^2 \mid x_1 + x_2 = 1\}.$$

Here the weighted sum problem has a minimal solution only for the weights $w_1 = w_2 = 0.5$. For all other parameters there are feasible points but the scalar problem is not solvable.

We further want to mention the weighted p-power method ([149]) where the scalarization is given by

$$\min_{x \in \Omega} \sum_{i=1}^{m} w_i f_i^p(x) \tag{2.39}$$

for $p \geq 1$, $w \in \mathbb{R}_+^m \setminus \{0_m\}$ (or $w \in K^* \setminus \{0_m\}$). For $p = 1$ (2.39) is equal to the weighted sum method. For arbitrary p the problem (2.39) can be seen as an application of the weighted sum method to the multiobjective optimization problem

$$\min_{x \in \Omega} \begin{pmatrix} f_1^p(x) \\ \vdots \\ f_m^p(x) \end{pmatrix}.$$

Another generalization of the weighted sum method is discussed in [231, pp.111f]. There k ($k \in \mathbb{N}$, $k \leq m$) linearly independent weights $w^1, \ldots, w^k \in \mathbb{R}_+^m$ are allowed representing for instance the preferences of k decision makers. Besides a reference point $v \in \mathbb{R}^k$ is given. Then the problem

$$\min_{x \in \Omega} \max_{i \in \{1,\ldots,k\}} (w^i)^\top f(x) - v_i \tag{2.40}$$

is solved. The connection to the parameters of the Pascoletti-Serafini problem is given by the equations

$$\begin{aligned} (w^i)^\top a &= v_i, & i &= 1, \ldots, k, \\ (w^i)^\top r &= 1, & i &= 1, \ldots, k, \end{aligned}$$

and $K := \{y \in \mathbb{R}^m \mid (w^i)^\top y \geq 0, \ i = 1, \ldots, k\}$. The set K is a closed convex cone and K is pointed if and only if $k = m$. A minimal solution of (2.40) is not only weakly K-minimal but because of $\mathbb{R}_+^m \subset K$ also weakly EP-minimal ([231, Theorem 7.2.1a)]). For $k = 1$ and $v = 0$ (2.40) is equivalent to (2.37). In the following section we will discuss a special case of the problem (2.40) for $k = m$.

2.5.8 Problem According to Kaliszewski

In [134] Kaliszewski discusses the following problem called (P^∞):

$$\min_{x \in \Omega} \max_{i \in \{1,\ldots,m\}} \lambda_i \left((f_i(x) - y_i^*) + \rho \sum_{j=1}^{m} (f_j(x) - y_j^*) \right) \quad (2.41)$$

for a closed set $f(\Omega) \subset y^* + \text{int}(\mathbb{R}^m_+)$, $y^* \in \mathbb{R}^m$, and $\rho > 0$, $\lambda_i > 0$, $i = 1, \ldots, m$, for determining properly efficient solutions ([134, Theorem 4.2]). The connection to the Pascoletti-Serafini problem is given by the following theorem (compare [231, pp.118f]):

Theorem 2.42. *A point $\bar{x}$ is a minimal solution of (P^∞) for $f(\Omega) \subset y^* + \text{int}(\mathbb{R}^m_+)$, $y^* \in \mathbb{R}^m$, $\rho > 0$, and $\lambda_i > 0$, $i = 1, \ldots, m$, if and only if the point $(\bar{t}, \bar{x})$ with*

$$\bar{t} = \max_{i \in \{1,\ldots,m\}} \lambda_i \left((f_i(\bar{x}) - y_i^*) + \rho \sum_{j=1}^{m} (f_j(\bar{x}) - y_j^*) \right) \quad (2.42)$$

is a minimal solution of $(SP(a,r))$ with $a = y^*$, $r \in \mathbb{R}^m$ with

$$r_i + \rho \sum_{j=1}^{m} r_j = \frac{1}{\lambda_i}, \quad \text{for all } i = 1, \ldots, m, \quad (2.43)$$

and $K = \{y \in \mathbb{R}^m \mid y_i + \rho \sum_{i=1}^{m} y_j \geq 0, \ i = 1, \ldots, m\}$.

Proof. For the parameter a and the cone K as in the theorem the problem $(SP(a,r))$ reads as follows

$$\min t$$

subject to the constraints

$$y_i^* + t\, r_i - f_i(x) + \rho \sum_{j=1}^{m} (y_j^* + t\, r_j - f_j(x)) \geq 0, \quad i = 1, \ldots, m,$$

$$t \in \mathbb{R}, \ x \in \Omega,$$

which is because of $r_i + \rho \sum_{j=1}^{m} r_j = \frac{1}{\lambda_i} > 0$ equivalent to

$$\min t$$

subject to the constraints

$$t \geq \frac{f_i(x) - y_i^* + \rho \sum_{j=1}^{m} (f_j(x) - y_j^*)}{r_i + \rho \sum_{j=1}^{m} r_j}, \quad i = 1, \ldots, m,$$

$$t \in \mathbb{R}, \ x \in \Omega.$$

Using (2.43) a point $(\bar{t}, \bar{x})$ is a minimal solution of this problem if and only if $\bar{x}$ is a solution of (2.41) with $\bar{t}$ as in (2.42). □

The set K is a closed pointed convex cone and we have $r \in K$. For $m = 2$ the cone K is given by the set

$$K = \left\{ y \in \mathbb{R}^2 \;\middle|\; \begin{pmatrix} 1+\rho & \rho \\ \rho & 1+\rho \end{pmatrix} y \geq 0_2 \right\}.$$

Hence the parameter ρ controls the cone K. A variation of the parameters λ and ρ lead to a variation of the parameter r while the parameter a is chosen constant as y^*. Because of $\mathbb{R}_+^m \subset K$ and $r \in K$ for $\lambda_i > 0$, $i = 1\ldots, m$, we have $\mathcal{E}_w(f(\Omega), K) \subset \mathcal{E}_w(f(\Omega), \mathbb{R}_+^m)$ and thus a minimal solution of (P^∞) is an at least weakly EP-minimal point of (MOP) as a result of Lemma 1.7.

2.5.9 Further Scalarizations

We have shown that many scalarization problems can be seen as a special case of the Pascoletti-Serafini method and hence that the results for the general problem can be applied to these special cases, too. The enumeration of special cases is not complete. For example in [231] a problem called hyperbola efficiency going back to [75] is discussed. However for a connection to the Pascoletti-Serafini problem K has to be defined as a convex set which is not a cone. Also a generalization of the weighted Chebyshev norm problem is mentioned there which can be connected to the Pascoletti-Serafini problem then using a closed pointed convex cone K.

There are many other scalarization approaches, too, which cannot be subsumed under the Pascoletti-Serafini method like the hybrid method ([62, p.101]), the elastic constraint method ([62, p.102]), or Benson's method ([11]). Literature with surveys about different scalarization approaches is listed in the introduction of this chapter.

3
Sensitivity Results for the Scalarizations

In this chapter we study the connection between the choice of the parameters a and r and the minimal solutions $(t(a,r), x(a,r))$ of the scalar optimization problem (SP(a,r)). Thereby we are especially interested in the generated weakly efficient points $f(x(a,r))$ of the multiobjective optimization problem (MOP). We still use the Assumption 2.25 and we recall the problem (SP(a,r)):

$$\text{(SP(a,r))} \quad \min_{t,x} \tilde{f}(t,x,a,r) := t$$

subject to the constraints
$$\tilde{g}^1(t,x,a,r) := a + tr - f(x) \in K,$$
$$\tilde{g}^2(t,x,a,r) := g(x) \in C,$$
$$\tilde{h}(t,x,a,r) := h(x) = 0_q,$$
$$t \in \mathbb{R}, \ x \in S.$$

For our considerations we define a minimal value function $\tau \colon \mathbb{R}^m \times \mathbb{R}^m \to \overline{\mathbb{R}} = \mathbb{R} \cup \{-\infty, +\infty\}$ to the parameter dependent scalar-valued optimization problem (SP(a,r)) by

$$\tau(a,r) := \inf\{t \in \mathbb{R} \mid (t,x) \in \Sigma(a,r)\}.$$

Here, $\Sigma(a,r)$ denotes the constraint set of the problem (SP(a,r)) dependent on a and r:

$$\Sigma(a,r) := \{(t,x) \in \mathbb{R} \times S \mid a+tr-f(x) \in K, \ g(x) \in C, \ h(x) = 0_q\}$$
$$= \{(t,x) \in \mathbb{R}^{n+1} \mid a+tr-f(x) \in K, \ x \in \Omega\}.$$

We start our study with a reference problem (SP(a^0, r^0)) and we restrict ourselves to local investigations. Let (t^0, x^0) be a minimal solution of the reference problem. Then we are interested in the question how small changes of the parameters a and r influence the minimal value $t(a,r)$ and the minimal solutions $(t(a,r), x(a,r))$, and thus $f(x(a,r))$, locally in a neighborhood of (t^0, x^0). For that we consider for a $\delta > 0$ the local minimal value function $\tau^\delta \colon \mathbb{R}^m \times \mathbb{R}^m \to \overline{\mathbb{R}}$ defined by

$$\tau^\delta(a,r) := \inf\{t \in \mathbb{R} \mid (t,x) \in \Sigma(a,r) \cap B_\delta(t^0, x^0)\}$$

with $B_\delta(t^0, x^0)$ the closed ball with radius δ around the point x^0. We will see that under certain assumptions this function is differentiable and its derivative can be expressed with the help of the derivative of the Lagrange function to the problem $(\mathrm{SP}(a,r))$ w.r.t. the parameters. Based on this we approximate the local minimal value function. As a result we can locally predict the position of the points $f(x(a,r))$ depending on a change of the parameters a and r. Some examinations of the continuity and differentiability of the minimal value function of this scalarization approach can also be found in [213].

By knowing the connection between the parameters (a,r) and the weakly efficient points $f(x(a,r))$ we can approximate the efficient set in a neighborhood of $f(x^0)$. With that information we can determine further approximation points of the efficient set which have a controlled distance from the point $f(x^0)$.

3.1 Sensitivity Results in Partially Ordered Spaces

We first discuss the general case that the partial orderings appearing in the problem $(\mathrm{SP}(a,r))$ are induced by arbitrary closed pointed convex cones K and C. As the special case of the natural ordering, i.e. $K = \mathbb{R}^m_+$ and $C = \mathbb{R}^p_+$, is very interesting and allows some stronger results, we discuss this case in the following section in detail.

The differentiability of the local minimal value function of scalar-valued parametric optimization problems over normed linear spaces under certain assumptions was shown by Alt in [6]. We first present this general result by Alt before we apply it to our special problem. We need the notion of Fréchet-differentiability which we recall for convenience.

Definition 3.1. Let $(X, \|\cdot\|_X)$, $(Y, \|\cdot\|_Y)$ be normed linear spaces and let $\hat{S}$ be an open nonempty subset of X. Let the map $F \colon \hat{S} \to Y$ be given and $\bar{x} \in \hat{S}$. If there is a continuous linear map $F_x(\bar{x}) \colon X \to Y$ with the property

3.1 Sensitivity Results in Partially Ordered Spaces

$$\lim_{\|h\|_X \to 0} \frac{\|F(\bar{x}+h) - F(\bar{x}) - F_x(\bar{x})(h)\|_Y}{\|h\|_X} = 0,$$

then $F_x(\bar{x})$ is called Fréchet-derivative of F in $\bar{x}$ and F is called Fréchet-differentiable in $\bar{x}$.

If $F \colon \mathbb{R}^n \to \mathbb{R}$ is a real-valued differentiable function then the Fréchet-derivative of F in a point $\bar{x} \in \mathbb{R}^n$ equals the gradient of F in $\bar{x}$. Hence it is

$$F_x(\bar{x})(h) = \nabla F(\bar{x})^\top h$$

for $h \in \mathbb{R}^n$.

The notions of the Lagrange function and the Lagrange multipliers (compare p.47) can be generalized to arbitrary normed linear spaces using the Fréchet-derivative (see e.g. [123, p.119]). Alt ([6, Theorem 5.3, 6.1]) shows that – under certain assumptions – the derivative of the local minimal value function of a parameter dependent optimization problem equals the derivative of the Lagrange function w.r.t. the parameters:

Theorem 3.2. *Let X, Y, Z be Banach spaces, W a normed linear space, $\hat{S}$ a nonempty subset of X, $S \subset \hat{S}$ a closed convex set and $C \subset Y$ a closed convex cone. Further let the maps $F \colon \hat{S} \times W \to \mathbb{R}$, $G \colon \hat{S} \times W \to Y$, and $H \colon \hat{S} \times W \to Z$ be given.*

For $w \in W$ we consider the parameter dependent optimization problem

$$(P(w)) \qquad \min F(x,w)$$

subject to the constraints

$$G(x,w) \in C,$$
$$H(x,w) = 0_Z,$$
$$x \in S.$$

The constraint set depending on the parameter $w \in W$ is then given by

$$\Sigma(w) := \{x \in S \mid G(x,w) \in C, \ H(x,w) = 0_Z\}.$$

Let $(P(w^0))$ be the so-called reference problem and let the point x^0 be a local minimal solution of $(P(w^0))$. Assume the following:

a) The set $N_1(x^0) \subset \hat{S}$ is a neighborhood of x^0 and $N_1(w^0) \subset W$ of w^0, such that the maps $F(\cdot, w)$, $G(\cdot, w)$, and $H(\cdot, w)$ are twice Fréchet-differentiable on $N_1(x^0)$ for all $w \in N_1(w^0)$.

70   3 Sensitivity Results for the Scalarizations

b) The maps G, H, F_x, G_x, H_x, F_{xx}, G_{xx}, and H_{xx} are continuous on $N_1(x^0) \times N_1(w^0)$.
c) There is a neighborhood $N_2(x^0) \subset N_1(x^0)$ of x^0 and a neighborhood $N_2(w^0) \subset N_1(w^0)$ of w^0, such that F_x, G, G_x, H, and H_x satisfy the following Lipschitz condition for all $x \in N_2(x^0)$ and for all $w^1, w^2 \in N_2(w^0)$ with constants c'_F, c_G, c'_G, c_H, and c'_H:

$$\begin{aligned}
\|F_x(x, w^1) - F_x(x, w^2)\| &\leq c'_F \|w^1 - w^2\|, \\
\|G(x, w^1) - G(x, w^2)\| &\leq c_G \|w^1 - w^2\|, \\
\|G_x(x, w^1) - G_x(x, w^2)\| &\leq c'_G \|w^1 - w^2\|, \\
\|H(x, w^1) - H(x, w^2)\| &\leq c_H \|w^1 - w^2\|, \\
\|H_x(x, w^1) - H_x(x, w^2)\| &\leq c'_H \|w^1 - w^2\|.
\end{aligned}$$

d) The maps F, G, and H are continuously Fréchet-differentiable on $N_2(x^0) \times N_2(w^0)$.
e) The point x^0 is regular for the set $\Sigma(w^0)$, i. e.

$$0_{Y \times Z} \in \text{int}\left\{\begin{pmatrix} G(x^0, w^0) \\ 0_Z \end{pmatrix} + \begin{pmatrix} G_x(x^0, w^0)(x - x^0) \\ H_x(x^0, w^0)(x - x^0) \end{pmatrix} - \begin{pmatrix} c \\ 0_Z \end{pmatrix} \Big| \right.$$
$$\left. x \in S, \ c \in C \right\}.$$

f) The following strict second order sufficient condition is satisfied: there are Lagrange multipliers $(\mu^0, \xi^0) \in C^* \times Z^*$ and a constant $\alpha > 0$, such that for the second Fréchet-derivative of the Lagrange function $\mathcal{L}(x, \mu, \xi, w)$ in (x^0, μ^0, ξ^0, w^0) it is

$$\mathcal{L}_{xx}(x^0, \mu^0, \xi^0, w^0)(x, x) \geq \alpha \|x\|^2$$

for all $x \in X$ with $H_x(x^0, w^0)(x) = 0_Z$.
g) For $P := X^* \times Y \times Z$ let B_P denote the closed unit ball in P. Then there exists a $\zeta > 0$ such that for $p^1, p^2 \in \zeta B_P$ arbitrarily chosen with $p^i = (x^{*i}, u^i, v^i)$, $i = 1, 2$, it holds:
If x^1 and x^2 respectively are solutions of the quadratic optimization problem $(QP)_{p^i}$ for $i = 1, 2$ given by

$(QP)_{p^i}$
$$\min J(x, p^i)$$
subject to the constraints
$$G(x^0, w^0) + G_x(x^0, w^0)(x - x^0) - u^i \in C,$$
$$H_x(x^0, w^0)(x - x^0) - v^i = 0_Z,$$
$$x \in S$$

3.1 Sensitivity Results in Partially Ordered Spaces

with

$$J(x, p^i) := \frac{1}{2}\mathcal{L}_{xx}(x^0, \mu^0, \xi^0, w^0)(x - x^0, x - x^0)$$
$$+ F_x(x^0, w^0)(x - x^0) - x^{*i}(x - x^0),$$

then the Lagrange multipliers (μ^i, ξ^i) to the point x^i, $i = 1, 2$, are unique and there is a constant c_M with

$$\|(\mu^1, \xi^1) - (\mu^2, \xi^2)\| \leq c_M \left(\|x^1 - x^2\| + \|p^1 - p^2\|\right).$$

Then there exists a constant $\delta > 0$ and a neighborhood $N(w^0) \subset W$ of w^0, such that the local minimal value function $\tau^\delta \colon W \to \mathbb{R}$,

$$\tau^\delta(w) := \inf\{F(x, w) \in \mathbb{R} \mid x \in \Sigma(w) \cap B_\delta(x^0)\},$$

is Fréchet-differentiable on $N(w^0)$ with Fréchet-derivative

$$\tau^\delta_w(w) = \mathcal{L}_w(x(w), \mu(w), \xi(w), w).$$

Here $x(w)$ denotes the local unique minimal solution of $(P(w))$ and $(\mu(w), \xi(w))$ are the unique Lagrange multipliers to the point $x(w)$. Besides the map $\phi \colon N(w^0) \to B_\delta(x^0) \times B_\delta(\mu^0, \xi^0)$ with

$$\phi(w) := (x(w), \mu(w), \xi(w))$$

is Lipschitz continuous on $N(w^0)$.

The proof in [6] uses an implicit function theorem by Robinson ([187]). The preceding theorem delivers the connection between the parameters and the minimal value and the minimal solutions of a parameter dependent optimization problem.

We will not apply Theorem 3.2 directly to the problem $(SP(a, r))$ but to the modified problem $(\overline{SP}(a, r))$. As the following example shows a direct examination of the problem $(SP(a, r))$ has a disadvantage: we cannot conclude from the dependence of the minimal value $t(a, r)$ on a and r how a change of the parameters influences the generated weakly efficient points $f(x(a, r))$.

Example 3.3. We consider the bicriteria optimization problem

3 Sensitivity Results for the Scalarizations

$$\min \begin{pmatrix} x_1 \\ x_2 \end{pmatrix}$$

subject to the constraints
$$1 \leq x_1 \leq 3,$$
$$1 \leq x_2 \leq 3,$$
$$x \in \mathbb{R}^2$$

with $K = \mathbb{R}_+^2$, see Fig. 3.1.

The correspondent scalarization (SP(a,r)) is given by

$$\min t$$

subject to the constraints
$$a + tr - f(x) \geq_2 0_2,$$
$$x \in \Omega, \ t \in \mathbb{R}.$$

Let the problem (SP(a^0, r^0)) with

$$a^0 = \begin{pmatrix} 2 \\ 5/2 \end{pmatrix}, \ r^0 = \begin{pmatrix} 1/2 \\ 1/4 \end{pmatrix}$$

denote the reference problem. The point (t^0, x^0) with $t^0 = -2$, $x^0 = (1, \frac{3}{2})$ is a minimal solution of the reference problem. We set the parameter r^0 constant and vary the parameter a^0 by moving it in direction

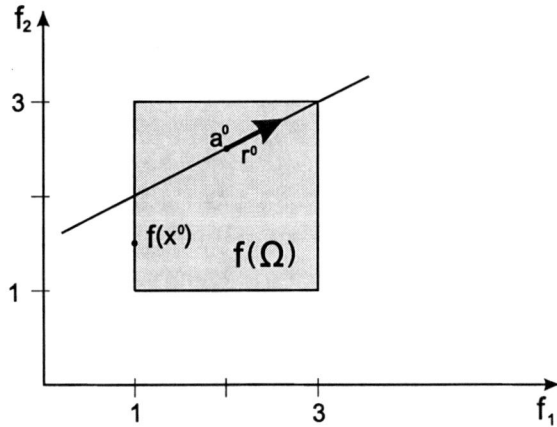

Fig. 3.1. Visualization of Example 3.3.

3.1 Sensitivity Results in Partially Ordered Spaces

$(1,0)$, i.e. we consider the problem $(\text{SP}(a^\varepsilon, r^0))$ with $a^\varepsilon := a^0 + \varepsilon \cdot (1,0)^\top$. For $0 \leq \varepsilon \leq 2$ we get as minimal value $t(\varepsilon) = -2 - 2\varepsilon$ dependent on the choice of ε. As for $0 < \varepsilon \leq 1$ the point $(t(\varepsilon), x^0)$ is still a minimal solution of $(\text{SP}(a^\varepsilon, r^0))$ we get the same weakly efficient point $f(x(\varepsilon)) = f(x^0)$ of the multiobjective optimization problem. Here, we cannot deduce a change of the points $f(x(\varepsilon))$ from a change of the minimal value $t(\varepsilon)$ (caused by a variation of the parameter a).

For $\varepsilon = 1$ the inequality constraint $a^\varepsilon + t r^0 - f(x) \geq_2 0_2$ is active in the point $(t(\varepsilon), x^0)$, i.e. $a^1 + t(1) r^0 = f(x^0)$. Then, a change in ε with $\varepsilon > 1$ results in a new weakly EP-minimal point $x(\varepsilon) \neq x^0$.

We cannot apply Theorem 3.2 directly to the problem $(\text{SP}(a,r))$ for a technical reason, too, as explained on p.74.

We have seen that for our considerations the sensitivity results about the minimal value are only useful applied to a reference problem $(\text{SP}(a^0, r^0))$ with the constraint $a^0 + t r^0 - f(x) \in K$ being active in (t^0, x^0), i.e. $a^0 + t^0 r^0 - f(x^0) = 0_m$. This is no difficulty because according to Theorem 2.21 for any minimal solution (t^0, x^0) of $(\text{SP}(a^0, r^0))$ there exists a parameter a' and some $t' \in \mathbb{R}$ so that (t', x^0) is a minimal solution of $(\text{SP}(a', r^0))$ with $a' + t' r^0 - f(x^0) = 0_m$. Thus, for being able to deduce changes of the point $f(x(a,r))$ from changes in the minimal value $t(a,r)$ depending on the parameters, we consider the modified problem $(\overline{SP}(a,r))$ instead of $(\text{SP}(a,r))$. We recall $(\overline{SP}(a,r))$:

$$\min t$$
subject to the constraints
$$(\overline{SP}(a,r)) \quad a + tr - f(x) = 0_m,$$
$$g(x) \in C,$$
$$h(x) = 0_q,$$
$$t \in \mathbb{R},\ x \in S.$$

Let

$$\overline{\Sigma}(a,r) := \{(t,x) \in \mathbb{R} \times S \mid a + tr - f(x) = 0_m,\ g(x) \in C,\ h(x) = 0_q\}$$
$$= \{(t,x) \in \mathbb{R}^{n+1} \mid a + tr - f(x) = 0_m,\ x \in \Omega\}$$

denote the constraint set. This modified problem has no longer the property, that for any minimal solution $(\bar{t}, \bar{x})$ the point $f(\bar{x})$ is an at least weakly efficient point of the multiobjective optimization problem (MOP) as already discussed in Sect. 2.4.

3 Sensitivity Results for the Scalarizations

Pascoletti and Serafini ([181]) have discussed the sensitivity of the problem (SP(a,r)), too. They have restricted themselves to the special case of a polyhedral cone K, $C = \mathbb{R}^p_+$ and to $S = \mathbb{R}^n$. The advantage of polyhedral cones – as already discussed in Sect. 1.2 – is, that they can be expressed by

$$K = \{x \in \mathbb{R}^m \mid x = \overline{K}\, y \text{ for some } y \in \mathbb{R}^s_+\}$$

with a matrix $\overline{K} \in \mathbb{R}^{m \times s}$. By introducing an additional variable $y \in \mathbb{R}^s$, the constraint $a + t\,r - f(x) \in K$ in (SP(a,r)) can be replaced by

$$\begin{aligned} a + t\,r - f(x) - \overline{K}\,y &= 0_m, \\ y &\geq_s 0_s. \end{aligned} \quad (3.1)$$

The resulting optimization problem has equality and inequality constraints w.r.t. the natural ordering only. Then, the constraints which are active or inactive in the reference problem can be examined separately. In the case of non-degeneracy it can be shown under some additional assumptions that the index set of the active inequality constraints is non-varying for small parameter changes. These active constraints can thus be handled as equality constraints. By applying the implicit function theorem further results follow. Using this approach Pascoletti and Serafini show that the minimal solution $x(a,r)$ is a function of the parameters a and r which is locally differentiable ([181, Theorem 4.1]). The differentiability and the derivative of the local minimal value function are not considered. We discuss this special case of a polyhedral cone again at the end of this section.

We examine again the problem (SP(a,r)) with a general closed convex cone K. The Hessian of the Lagrange function is given by

$$\nabla^2_{(t,x)} \mathcal{L}(t, x, \mu, \nu, \xi, a, r) = \begin{pmatrix} 0 & 0 \\ 0 & W(x, \mu, \nu, \xi) \end{pmatrix}$$

with

$$W(x, \mu, \nu, \xi) = \sum_{i=1}^{m} \mu_i \nabla^2_x f_i(x) - \sum_{j=1}^{p} \nu_j \nabla^2_x g_j(x) - \sum_{k=1}^{q} \xi_k \nabla^2_x h_k(x).$$

(The Lagrange function to the problem (SP(a,r)) has been introduced on p.48.) For all points $(\bar{t}, \bar{x}) = (\bar{t}, 0_n)$ with $\bar{t} \neq 0$ it is

3.1 Sensitivity Results in Partially Ordered Spaces   75

$$\nabla_{(t,x)}\tilde{h}(t,x,a,r)\begin{pmatrix}\bar{t}\\0_n\end{pmatrix} = (0_n, \nabla_x h(x))\begin{pmatrix}\bar{t}\\0_n\end{pmatrix} = 0_q$$

and

$$(\bar{t}, 0_n^\top)\nabla^2_{(t,x)}\mathcal{L}(t,x,\mu,\nu,\xi,a,r)\begin{pmatrix}\bar{t}\\0_n\end{pmatrix} = 0.$$

However we have

$$\alpha\left\|\begin{pmatrix}\bar{t}\\0_n\end{pmatrix}\right\|^2 = \alpha\,|\bar{t}|^2 > 0$$

for $\alpha > 0$ contradicting the assumption f) of Theorem 3.2.

Therefore we cannot apply Theorem 3.2 to the problem (SP(a,r)) and we turn our attention to the modified problem ($\overline{\text{SP}}(a,r)$). This has further the advantage of delivering a connection between the minimal value $t(a,r)$ and the points $f(x(a,r))$. We start by discussing whether Theorem 3.2 is applicable to the modified problem. We need the following assumption:

Assumption 3.4. Let the Assumption 2.25 hold. Further let the functions f, g, and h be twice continuously differentiable on $\hat{S}$.

Remark 3.5. A consequence of Assumption 3.4 is that the functions $\tilde{f}$, $\tilde{g}^1$, $\tilde{g}^2$, and $\tilde{h}$ of the optimization problem (SP(a,r)) are twice continuously differentiable on $\mathbb{R} \times \hat{S} \times \mathbb{R}^m \times \mathbb{R}^m$.

Lemma 3.6. *Let Assumption 3.4 hold. Let (t^0, x^0) be a local minimal solution of $(\overline{\text{SP}}(a^0, r^0))$ with Lagrange multipliers $(\mu^0, \nu^0, \xi^0) \in \mathbb{R}^m \times C^* \times \mathbb{R}^q$. Assume there exists a constant $\tilde{\alpha} > 0$ such that for the matrix*

$$W(x^0, \mu^0, \nu^0, \xi^0) = \mu^{0\top}\nabla^2_x f(x^0) - \nu^{0\top}\nabla^2_x g(x^0) - \xi^{0\top}\nabla^2_x h(x^0)$$

we have

$$x^\top W(x^0, \mu^0, \nu^0, \xi^0)x \geq \tilde{\alpha}\,\|x\|^2 \tag{3.2}$$

for all $x \in \{x \in \mathbb{R}^n \mid \nabla_x h(x^0)x = 0_q,\ \nabla_x f(x^0)x = r^0 t\ \text{for a}\ t \in \mathbb{R}\}$. Then there is a constant $\alpha > 0$ such that for the Lagrange function $\overline{\mathcal{L}}$ to $(\overline{\text{SP}}(a,r))$ we have

$$(t, x^\top)\nabla^2_{(t,x)}\overline{\mathcal{L}}(t^0, x^0, \mu^0, \nu^0, \xi^0, a^0, r^0)\begin{pmatrix}t\\x\end{pmatrix} \geq \alpha\left\|\begin{pmatrix}t\\x\end{pmatrix}\right\|^2 \tag{3.3}$$

for all $(t,x) \in \{(t,x) \in \mathbb{R}\times\mathbb{R}^n \mid \nabla_x h(x^0)x = 0_q,\ \nabla_x f(x^0)x = r^0 t\}$, i.e. assumption f) of Theorem 3.2 is satisfied for the problem $(\overline{\text{SP}}(a^0, r^0))$.

3 Sensitivity Results for the Scalarizations

Proof. Because (t^0, x^0) is a local minimal solution of $(\overline{\text{SP}}(a^0, r^0))$ with Lagrange multipliers (μ^0, ν^0, ξ^0) we have for the associated Lagrange function

$$\nabla_{(t,x)} \overline{\mathcal{L}}(t^0, x^0, \mu^0, \nu^0, \xi^0, a^0, r^0)^\top \begin{pmatrix} t - t^0 \\ x - x^0 \end{pmatrix} \geq 0 \quad \text{for all } t \in \mathbb{R}, \ x \in S. \tag{3.4}$$

With

$$\frac{\partial \overline{\mathcal{L}}(t^0, x^0, \mu^0, \nu^0, \xi^0, a^0, r^0)}{\partial t} = 1 - \mu^{0\top} r^0$$

and because (3.4) has to be fulfilled for all $t \in \mathbb{R}$ we have

$$\mu^{0\top} r^0 = 1 \tag{3.5}$$

and so $\mu^0 \neq 0_m$, $r^0 \neq 0_m$. Because in $\mathbb{R}^n$ and $\mathbb{R}^m$ respectively all norms are equivalent, for all $(t, x) \in \mathbb{R} \times \mathbb{R}^n$ there exist positive constants $M^l, M^u \in \mathbb{R}_+$ and $\tilde{M}^l, \tilde{M}^u \in \mathbb{R}_+$ respectively with

$$M^l \|x\|_2 \leq \|x\| \leq M^u \|x\|_2$$

and

$$\tilde{M}^l \left\| \begin{pmatrix} t \\ x \end{pmatrix} \right\|_2 \leq \left\| \begin{pmatrix} t \\ x \end{pmatrix} \right\| \leq \tilde{M}^u \left\| \begin{pmatrix} t \\ x \end{pmatrix} \right\|_2.$$

(For instance for the Euclidean norm $\|\cdot\| = \|\cdot\|_2$ this is true for $M^l = M^u = 1$ and $\tilde{M}^l = \tilde{M}^u = 1$ respectively.) For all $(t, x) \in \mathbb{R} \times \mathbb{R}^n$ with $\nabla f_x(x^0) x = r^0 t$ we have together with (3.5) the equation

$$\mu^{0\top} \nabla_x f(x^0) x = t$$

and then we get the upper bound

$$|t|^2 = |\mu^{0\top} \nabla_x f(x^0) x|^2 \leq \|\mu^0\|_2^2 \|\nabla_x f(x^0)\|_2^2 \|x\|_2^2.$$

If we set now

$$\alpha := \frac{\tilde{\alpha} (M^l)^2}{(\tilde{M}^u)^2 \left(1 + \|\mu^0\|_2^2 \|\nabla_x f(x^0)\|_2^2\right)} > 0$$

we conclude from (3.2) for all $(t, x) \in \{(t, x) \in \mathbb{R} \times \mathbb{R}^n \mid \nabla_x h(x^0) x = 0_q, \ \nabla_x f(x^0) x = r^0 t\}$

$$x^\top W(x^0, \mu^0, \nu^0, \xi^0) x \geq \tilde{\alpha} \|x\|^2$$

3.1 Sensitivity Results in Partially Ordered Spaces

$$\geq \tilde{a}\,(M^l)^2 \|x\|_2^2$$
$$= \alpha\,(\tilde{M}^u)^2\,\left(1 + \|\mu^0\|_2^2\,\|\nabla_x f(x^0)\|_2^2\right)\,\|x\|_2^2$$
$$\geq \alpha\,(\tilde{M}^u)^2\,(\|x\|_2^2 + |t|^2)$$
$$= \alpha\,(\tilde{M}^u)^2\,\left\|\begin{pmatrix} t \\ x \end{pmatrix}\right\|_2^2$$
$$\geq \alpha\,\left\|\begin{pmatrix} t \\ x \end{pmatrix}\right\|^2.$$

With

$$\nabla^2_{(t,x)}\overline{\mathcal{L}}(t^0, x^0, \mu^0, \nu^0, \xi^0, a^0, r^0) = \begin{pmatrix} 0 & 0 \\ 0 & W(x^0, \mu^0, \nu^0, \xi^0) \end{pmatrix}$$

the assertion is proven. □

The condition (3.3) for all (t, x) in which the derivatives of the equality constraints are equal to zero is called strict second order sufficient condition. If this condition is fulfilled for a regular point (see assumption e) of Theorem 3.2) then this is sufficient for strict local minimality of the considered point ([162, Theorem 5.2]).

We can now apply Theorem 3.2 to the problem $(\overline{SP}(a, r))$.

Theorem 3.7. *Let Assumption 3.4 and the assumptions of Lemma 3.6 hold. We consider the parameter dependent optimization problem $(\overline{SP}(a, r))$ with the constraint set $\overline{\Sigma}(a, r)$ starting with a reference problem $(\overline{SP}(a^0, r^0))$ with a local minimal solution (t^0, x^0) and with Lagrange multipliers $(\mu^0, \nu^0, \xi^0) \in \mathbb{R}^m \times C^* \times \mathbb{R}^q$.*

i) Suppose the point (t^0, x^0) is regular for the set $\overline{\Sigma}(a^0, r^0)$, i.e. $0_{m+p+q} \in$

$$\mathrm{int}\left\{\begin{pmatrix} 0_m \\ g(x^0) \\ 0_q \end{pmatrix} + \begin{pmatrix} r^0(t - t^0) - \nabla_x f(x^0)(x - x^0) \\ \nabla_x g(x^0)(x - x^0) \\ \nabla_x h(x^0)(x - x^0) \end{pmatrix} - \begin{pmatrix} 0_m \\ c \\ 0_q \end{pmatrix} \Bigg| \right.$$
$$\left. c \in C,\ x \in S,\ t \in \mathbb{R}\right\}.$$

ii) Assume there exists a $\zeta > 0$ such that for arbitrary points $p^1, p^2 \in \zeta\tilde{B}$ (with $\tilde{B}$ the closed unit ball in $\mathbb{R}^{1+n+m+p+q}$) with $p^i = (t^{*i}, x^{*i}, u^i, v^i, w^i)$, $i = 1, 2$, it holds: if (t^1, x^1) and (t^2, x^2) respectively are solutions of the quadratic optimization problem determined by

$$\min J(t, x, p^i)$$

subject to the constraints

$$r^0 (t - t^0) - \nabla_x f(x^0)(x - x^0) - u^i = 0_m,$$
$$g(x^0) + \nabla_x g(x^0)(x - x^0) - v^i \in C,$$
$$\nabla_x h(x^0)(x - x^0) - w^i = 0_q,$$
$$t \in \mathbb{R}, \ x \in S,$$

$(i = 1, 2)$ with

$$J(t, x, p^i) := \frac{1}{2}(x-x^0)^\top W\,(x-x^0) + (t-t^0) - t^{*i}\,(t-t^0) - (x^{*i})^\top (x-x^0),$$

and

$$W := W(x^0, \mu^0, \nu^0, \xi^0) = \mu^{0\top}\nabla_x^2 f(x^0) - \nu^{0\top}\nabla_x^2 g(x^0) - \xi^{0\top}\nabla_x^2 h(x^0),$$

then the Lagrange multipliers (μ^i, ν^i, ξ^i) to the solutions (t^i, x^i), $i = 1, 2$, are unique and

$$\|(\mu^1, \nu^1, \xi^1) - (\mu^2, \nu^2, \xi^2)\| \leq c_M \left(\|(t^1, x^1) - (t^2, x^2)\| + \|p^1 - p^2\| \right)$$

with some constant c_M.

Then there exists a $\delta > 0$ and a neighborhood $N(a^0, r^0)$ of (a^0, r^0) such that the local minimal value function

$$\bar{\tau}^\delta(a, r) := \inf\{t \in \mathbb{R} \mid (t, x) \in \overline{\Sigma}(a, r) \cap B_\delta(t^0, x^0)\}$$

is differentiable on $N(a^0, r^0)$ with the derivative

$$\nabla_{(a,r)} \bar{\tau}^\delta(a, r) = \nabla_{(a,r)} \overline{\mathcal{L}}(\bar{t}(a, r), \bar{x}(a, r), \mu(a, r), \nu(a, r), \xi(a, r), a, r).$$

Here $(\bar{t}(a, r), \bar{x}(a, r))$ denotes the strict local minimal solution of $(\overline{\text{SP}}(a, r))$ for $(a, r) \in N(a^0, r^0)$ with the unique Lagrange multipliers $(\mu(a, r)$, $\nu(a, r), \xi(a, r))$. In addition to that the mapping

$$\phi \colon N(a^0, r^0) \to B_\delta(t^0, x^0) \times B_\delta(\mu^0, \nu^0, \xi^0)$$

defined by

$$\phi(a, r) := (\bar{t}(a, r), \bar{x}(a, r), \mu(a, r), \nu(a, r), \xi(a, r))$$

is Lipschitz continuous.

3.1 Sensitivity Results in Partially Ordered Spaces

Proof. We show that all assumptions of Theorem 3.2 are satisfied. According to Remark 3.5 the Assumption 3.4 implies that the assumptions a), b) and d) of Theorem 3.2 are fulfilled. According to Assumption 3.4 the functions $\nabla_{(t,x)}\tilde{f}$, $\tilde{g}^1$, $\nabla_{(t,x)}\tilde{g}^1$, $\tilde{g}^2$, $\nabla_{(t,x)}\tilde{g}^2$, $\tilde{h}$, and $\nabla_{(t,x)}\tilde{h}$ are partially differentiable w.r.t. (a,r) and hence a local Lipschitz condition w.r.t (a,r) is satisfied in a neighborhood of (a^0, r^0). Thus condition c) is fulfilled. The condition i) and ii) are the correspondent conditions to e) and g) respectively. By applying Lemma 3.6 we immediately conclude that assumption f) of Theorem 3.2 is satisfied and therefore all assumptions of Theorem 3.2 are met. □

Remark 3.8. The condition ii) of the preceding theorem and the condition g) of Theorem 3.2 respectively are always satisfied, if we have only equality constraints ([6, Theorem 7.1]), or, in the case of the natural ordering $C = \mathbb{R}^n_+$, if the gradients of the active constraints are linearly independent ([78, Theorem 2.1], [131, Theorem 2], [187, Theorem 4.1]).

Lemma 3.9. *Let the assumptions of Theorem 3.7 be fulfilled with $S = \mathbb{R}^n$. Then there is a $\delta > 0$ and a neighborhood $N(a^0, r^0)$ of (a^0, r^0) such that the derivative of the local minimal value function w.r.t. the parameter a is given by*

$$\nabla_a \bar{\tau}^\delta(a,r) = -\mu(a,r) - \nabla_a \nu(a,r)^\top g(\bar{x}(a,r))$$

and w.r.t. the parameter r by

$$\nabla_r \bar{\tau}^\delta(a,r) = -\bar{t}(a,r)\mu(a,r) - \nabla_r \nu(a,r)^\top g(\bar{x}(a,r))$$

for all $(a,r) \in N(a^0, r^0)$. Here $(\bar{t}(a,r), \bar{x}(a,r))$ denotes the strict local minimal solution of $(\overline{SP}(a,r))$ for $(a,r) \in N(a^0, r^0)$ with the unique Lagrange multipliers $(\mu(a,r), \nu(a,r), \xi(a,r))$.

Proof. According to Theorem 3.7 there is a neighborhood $N(a^0, r^0)$ of (a^0, r^0) such that for all $(a,r) \in N(a^0, r^0)$ there is a strict local minimal solution $(\bar{t}(a,r), \bar{x}(a,r))$ with unique Lagrange multipliers $(\mu(a,r), \nu(a,r), \xi(a,r))$. For the derivative of the Lagrange function we have because of $S = \mathbb{R}^n$

$$\nabla_{(t,x)}\overline{\mathcal{L}}(\bar{t}(a,r), \bar{x}(a,r), \mu(a,r), \nu(a,r), \xi(a,r), a, r) = 0_{n+1}.$$

Then it follows

80   3 Sensitivity Results for the Scalarizations

$$0_m = \nabla_a \begin{pmatrix} \bar{t}(a,r) \\ \bar{x}(a,r) \end{pmatrix}^\top \nabla_{(t,x)} \overline{\mathcal{L}}(\bar{t}(a,r), \bar{x}(a,r), \mu(a,r), \nu(a,r), \xi(a,r), a, r)$$

$$= \nabla_a \bar{t}(a,r) - \sum_{i=1}^{m} \mu_i(a,r) \left(\nabla_a \bar{t}(a,r) r_i - \nabla_a \bar{x}(a,r)^\top \nabla_x f_i(\bar{x}(a,r)) \right)$$

$$- \sum_{j=1}^{p} \nu_j(a,r) \nabla_a \bar{x}(a,r)^\top \nabla_x g_j(\bar{x}(a,r))$$

$$- \sum_{k=1}^{q} \xi_k(a,r) \nabla_a \bar{x}(a,r)^\top \nabla_x h_k(\bar{x}(a,r)). \qquad (3.6)$$

In the same way we get

$$0_m = \nabla_r \begin{pmatrix} \bar{t}(a,r) \\ \bar{x}(a,r) \end{pmatrix}^\top \nabla_{(t,x)} \overline{\mathcal{L}}(\bar{t}(a,r), \bar{x}(a,r), \mu(a,r), \nu(a,r), \xi(a,r), a, r)$$

$$= \nabla_r \bar{t}(a,r) - \sum_{i=1}^{m} \mu_i(a,r) \left(\nabla_r \bar{t}(a,r) r_i - \nabla_r \bar{x}(a,r)^\top \nabla_x f_i(\bar{x}(a,r)) \right)$$

$$- \sum_{j=1}^{p} \nu_j(a,r) \nabla_r \bar{x}(a,r)^\top \nabla_x g_j(\bar{x}(a,r))$$

$$- \sum_{k=1}^{q} \xi_k(a,r) \nabla_r \bar{x}(a,r)^\top \nabla_x h_k(\bar{x}(a,r)).$$

According to Theorem 3.7 there exists a $\delta > 0$ such that the derivative of the local minimal value function is given by

$$\nabla_{(a,r)} \overline{\tau}^\delta(a,r) = \nabla_{(a,r)} \overline{\mathcal{L}}(\bar{t}(a,r), \bar{x}(a,r), \mu(a,r), \nu(a,r), \xi(a,r), a, r).$$

Together with derivation rules and (3.6) we conclude

$$\nabla_a \overline{\tau}^\delta(a,r) = \nabla_a \bar{t}(a,r) - \sum_{i=1}^{m} \mu_i(a,r) \left(e_i + \nabla_a \bar{t}(a,r) r_i \right.$$

$$\left. - \nabla_a \bar{x}(a,r)^\top \nabla_x f_i(\bar{x}(a,r)) \right)$$

$$- \sum_{i=1}^{m} \nabla_a \mu_i(a,r) \underbrace{(a_i + \bar{t}(a,r) r_i - f_i(\bar{x}(a,r)))}_{=0}$$

3.1 Sensitivity Results in Partially Ordered Spaces

$$-\sum_{j=1}^{p}\nu_j(a,r)\nabla_a\bar{x}(a,r)^\top\nabla_x g_j(\bar{x}(a,r)) - \sum_{j=1}^{p}\nabla_a\nu_j(a,r)g_j(\bar{x}(a,r))$$

$$-\sum_{k=1}^{q}\xi_k(a,r)\nabla_a\bar{x}(a,r)^\top\nabla_x h_k(\bar{x}(a,r)) - \sum_{k=1}^{q}\nabla_a\xi_k(a,r)\underbrace{h_k(\bar{x}(a,r))}_{=0}$$

$$= -\mu(a,r) - \nabla_a\nu(a,r)^\top g(\bar{x}(a,r)).$$

By using

$$\nabla_r(a_i + \bar{t}(a,r)r_i - f_i(\bar{x}(a,r))) = \bar{t}(a,r)e_i + \nabla_r\bar{t}(a,r)r_i -$$

$$\nabla_r\bar{x}(a,r)^\top\nabla_x f_i(\bar{x}(a,r))$$

we get just as well

$$\nabla_r\bar{\tau}^\delta(a,r) = -\bar{t}(a,r)\mu(a,r) - \nabla_r\nu(a,r)^\top g(\bar{x}(a,r)).$$

□

In the case of $C = \mathbb{R}^p_+$ the inequality $g(x) \in C$ equals the inequalities $g_j(x) \geq 0$, $j = 1, \ldots, p$, and then we can differ between active constraints $g_j(x^0) = 0$ and inactive constraints $g_j(x^0) > 0$ in the point (t^0, x^0). Because of the continuity of g_j and $\bar{x}(a,r)$, inactive constraints remain inactive in a neighborhood $N(a^0, r^0)$ of (a^0, r^0) ([79, Theorem 3.2.2 and Proof of Theorem 3.4.1]). Hence, for the associated Lagrange multipliers we have $\nu_j(a,r) = 0$ for all $(a,r) \in N(a^0, r^0)$ and thus $\nabla_{(a,r)}\nu_j(a^0, r^0) = 0_{2m}$. As a consequence we have for $C = \mathbb{R}^p_+$

$$\nabla_{(a,r)}\nu(a^0,r^0)^\top g(\bar{x}(a^0,r^0)) = \sum_{j=1}^{p}\nabla_{(a,r)}\nu_j(a^0,r^0)g_j(\bar{x}(a^0,r^0)) = 0_{2m}$$

and it follows:

Corollary 3.10. *Under the assumptions of Lemma 3.9 and with $C = \mathbb{R}^p_+$ it is*

$$\nabla_{(a,r)}\bar{\tau}^\delta(a^0,r^0) = -\begin{pmatrix} \mu^0 \\ t^0\mu^0 \end{pmatrix}.$$

So the derivative of the local minimal value function (and with that of the function $t(\cdot,\cdot)$) w.r.t. the parameter a in the point (a^0, r^0) is just the negative of the Lagrange multiplier μ^0. We get this Lagrange multiplier to the constraint $a^0 + t\,r^0 - f(x) \in K$ without additional effort by solving the scalar-valued problem (SP(a^0, r^0)). Thus, if we solve the optimization problem (SP(a^0, r^0)) with minimal solution (t^0, x^0) and Lagrange multiplier (μ^0, ν^0, ξ^0) (assuming the constraint $a^0 + t^0 r^0 - f(x) \in K$ is active, i.e. $a^0 + t^0 r^0 - f(x^0) = 0_m$), we get the following local first-order Taylor-approximation for the minimal value $t(a, r)$

$$t(a, r) \approx t^0 - \mu^{0\top}(a - a^0) - t^0\,\mu^{0\top}(r - r^0).$$

For that we assume that all necessary assumptions are satisfied and we suppose that the local minimal value function is several times differentiable. Examinations under which conditions this is assured can be found in [7, 18], [19, Theorem 4.102, 4.139 and 4.142], [202, 203, 204] and especially for the non-degenerated case and the natural ordering also in [79, Theorem 3.4.1].

We are primarily interested in the points $f(\bar{x}(a, r))$ of the set $f(\Omega)$, which are approximation points of the efficient set. The dependence of $f(\bar{x}(a, r))$ on a (and r) delivers important trade-off information for the decision maker, i.e. information how the improvement of one objective function causes the deterioration of another competing objective function. The notion of trade-off is discussed in more detail in [135].

Corollary 3.11. *Let the assumptions of Theorem 3.7 hold. Then, in a neighborhood of* (a^0, r^0),

$$\nabla_a f_i(\bar{x}(a, r)) = e_i + \nabla_a \bar{\tau}^\delta(a, r)\,r_i$$

and

$$\nabla_r f_i(\bar{x}(a, r)) = \bar{t}(a, r)\,e_i + \nabla_r \bar{\tau}^\delta(a, r)\,r_i$$

for $i = 1, \ldots, m$. *If additionally the assumptions of Corollary 3.10 are satisfied, then*

$$\nabla_a f(\bar{x}(a^0, r^0)) = E_m - r^0\,(\mu^0)^\top \ \text{and}\ \nabla_r f(\bar{x}(a^0, r^0)) = t^0\,E_m - t^0\,r^0\,(\mu^0)^\top.$$

Here e_i *denotes the ith unit vector in* $\mathbb{R}^m$ *and* E_m *the* (m, m)-*unit-matrix in* $\mathbb{R}^{m \times m}$.

We now discuss the already on p. 74 and in Lemma 1.18 mentioned special case of orderings, which are defined by polyhedral cones. In that case, i.e. for

$$K = \{x \in \mathbb{R}^m \mid \overline{K}x \geq_u 0_u\}$$

and

$$C = \{x \in \mathbb{R}^p \mid \overline{C}x \geq_v 0_v\}$$

with $\overline{K} \in \mathbb{R}^{u \times m}$ and $\overline{C} \in \mathbb{R}^{v \times p}$, $u, v \in \mathbb{N}$, the multiobjective optimization problem (MOP) can be reformulated as follows:

$$\min \overline{K} f(x)$$
subject to the constraints
$$\overline{C} g(x) \in \mathbb{R}^v_+,$$
$$h(x) = 0_q,$$
$$x \in S.$$

Now we look for the minimal points w. r. t. the natural ordering in $\mathbb{R}^u$. The correspondent scalarization is then

$$\min t$$
subject to the constraints
$$a_i + t\, r_i - \overline{k}^i f(x) \geq 0, \ i = 1, \ldots, u,$$
$$\overline{c}^j g(x) \geq 0, \ j = 1, \ldots, v,$$
$$h(x) = 0_q,$$
$$t \in \mathbb{R}, \ x \in S$$

with $\overline{k}^i$ and $\overline{c}^j$ the row vectors of the matrix $\overline{K}$ and $\overline{C}$ respectively. Using this reformulation we have reduced the problem to the case of the natural ordering and the sensitivity considerations can be subsumed under the examinations of the following section. Another possible reformulation using the introduction of new variables is demonstrated in (3.1).

3.2 Sensitivity Results in Naturally Ordered Spaces

In the following we discuss the special case that the image space of the multiobjective optimization problem is ordered w. r. t. the natural ordering, i.e. we try to find (weakly) EP-minimal points. Thus the ordering cone K equals the positive orthant $\mathbb{R}^m_+$. Additionally let $C = \mathbb{R}^p_+$ and $S = \mathbb{R}^n$. We summarize this:

Assumption 3.12. Let $K = \mathbb{R}^m_+$ and $C = \mathbb{R}^p_+$. Further assume $S = \hat{S} = \mathbb{R}^n$ and let $f \colon \mathbb{R}^n \to \mathbb{R}^m$, $g \colon \mathbb{R}^n \to \mathbb{R}^p$, and $h \colon \mathbb{R}^n \to \mathbb{R}^q$ be given functions.

3 Sensitivity Results for the Scalarizations

Under these assumptions we can write the multiobjective optimization problem as follows:

$$\text{(MOP)} \quad \begin{aligned} &\min f(x) \\ &\text{subject to the constraints} \\ &g_j(x) \geq 0, \quad j = 1, \ldots, p, \\ &h_k(x) = 0, \quad k = 1, \ldots, q, \\ &x \in \mathbb{R}^n. \end{aligned}$$

The correspondent parameter dependent scalarization problem (SP(a,r)) is then

$$\text{(SP}(a,r)) \quad \begin{aligned} &\min t \\ &\text{subject to the constraints} \\ &a_i + t\, r_i - f_i(x) \geq 0, \quad i = 1, \ldots, m, \\ &g_j(x) \geq 0, \quad j = 1, \ldots, p, \\ &h_k(x) = 0, \quad k = 1, \ldots, q, \\ &t \in \mathbb{R},\ x \in \mathbb{R}^n \end{aligned}$$

with parameters $a \in \mathbb{R}^m$ and $r \in \mathbb{R}^m$. Then the condition g) of Theorem 3.2 is already satisfied if the gradients of the active constraints are linearly independent ([6, p.22]).

Let (t^0, x^0) be a minimal solution of the reference problem (SP(a^0, r^0)) with Lagrange multipliers (μ^0, ν^0, ξ^0). We introduce index sets for the active non-degenerated, active degenerated and inactive constraints. Thus we define the following disjoint sets to the index sets $I := \{1, \ldots, m\}$ and $J := \{1, \ldots, p\}$:

$$\begin{aligned} I^+ &:= \{i \in I \mid a_i^0 + t^0\, r_i^0 - f_i(x^0) = 0,\ \mu_i^0 > 0\}, \\ I^0 &:= \{i \in I \mid a_i^0 + t^0\, r_i^0 - f_i(x^0) = 0,\ \mu_i^0 = 0\}, \\ I^- &:= \{i \in I \mid a_i^0 + t^0\, r_i^0 - f_i(x^0) > 0,\ \mu_i^0 = 0\} \end{aligned} \quad (3.7)$$

and

$$\begin{aligned} J^+ &:= \{j \in J \mid g_j(x) = 0,\ \nu_j^0 > 0\}, \\ J^0 &:= \{j \in J \mid g_j(x) = 0,\ \nu_j^0 = 0\}, \\ J^- &:= \{j \in J \mid g_j(x) > 0,\ \nu_j^0 = 0\}. \end{aligned} \quad (3.8)$$

We have $I = I^+ \cup I^0 \cup I^-$ and $J = J^+ \cup J^0 \cup J^-$. The active constraints in the point (t^0, x^0) are hence

3.2 Sensitivity Results in Naturally Ordered Spaces

$$a_i^0 + t^0 r_i^0 - f_i(x^0) \geq 0 \text{ for } i \in I^+ \cup I^0,$$
$$g_j(x) \geq 0 \text{ for } j \in J^+ \cup J^0 \quad \text{and}$$
$$h_k(x) = 0 \text{ for } k \in \{1, \ldots, q\}.$$

The active constraints with Lagrange multipliers equal to zero are called degenerated. Here these are the inequality constraints with the indices $i \in I^0$ and $j \in J^0$. As under the assumptions of Theorem 3.13 active non-degenerated constraints remain active under small parameter changes, we can treat these inequality constraints as equality constraints. Further inactive constraints stay inactive and can therefore be ignored. Because of that it is possible and sufficient in the special case of the natural ordering to consider the scalarization (SP(a, r)) directly instead of the modification ($\overline{SP}(a, r)$) (see also [6, pp.21f]):

Theorem 3.13. *Let the Assumption 3.12 hold. We consider the scalar-valued optimization problem (SP(a, r)) starting from the reference problem (SP(a^0, r^0)). Let (t^0, x^0) be a local minimal solution of (SP(a^0, r^0)) with Lagrange multipliers (μ^0, ν^0, ξ^0). Let there exist a $\gamma > 0$, such that the functions f, g, and h are twice continuously differentiable on an open neighborhood of $B_\gamma(x^0)$. Let the index sets I^+, I^0, I^-, and J^+, J^0, J^- be defined as in (3.7) and (3.8).*

Assume the following:

a) The gradients w. r. t. (t, x) in the point (t^0, x^0) of the (in the point (t^0, x^0)) active constraints of the problem (SP(a^0, r^0)), i. e. the vectors

$$\begin{pmatrix} r_i^0 \\ -\nabla_x f_i(x^0) \end{pmatrix}, \; i \in I^+ \cup I^0, \quad \begin{pmatrix} 0 \\ \nabla_x g_j(x^0) \end{pmatrix}, \; j \in J^+ \cup J^0,$$

$$\begin{pmatrix} 0 \\ \nabla_x h_k(x^0) \end{pmatrix}, \; k = 1, \ldots, q,$$

are linearly independent.

b) There is a constant $\alpha > 0$, such that for the Hessian of the Lagrange function $\mathcal{L}$ in the point (t^0, x^0) it holds

$$(t, x^\top) \nabla^2_{(t,x)} \mathcal{L}(t^0, x^0, \mu^0, \nu^0, \xi^0, a^0, r^0) \begin{pmatrix} t \\ x \end{pmatrix} \geq \alpha \left\| \begin{pmatrix} t \\ x \end{pmatrix} \right\|^2$$

for all

$(t,x) \in \{(t,x) \in \mathbb{R}^{n+1} \mid r_i^0 t = \nabla_x f_i(x^0)^\top x, \quad \forall\, i \in I^+,$

$$\nabla_x g_j(x^0)^\top x = 0 \quad \forall\, j \in J^+,$$

$$\nabla_x h_k(x^0)^\top x = 0 \quad \forall\, k = 1,\ldots,q\}.$$

Then the point (t^0, x^0) is a local unique minimal solution of the reference problem $(SP(a^0, r^0))$ with unique Lagrange multipliers (μ^0, ν^0, ξ^0) and there exists a $\delta > 0$ and a neighborhood $N(a^0, r^0)$ of (a^0, r^0), such that the local minimal value function $\tau^\delta : \mathbb{R}^m \times \mathbb{R}^m \to \overline{\mathbb{R}}$ with

$$\tau^\delta(a, r) := \inf\{t \in \mathbb{R} \mid (t, x) \in \Sigma(a, r) \cap B_\delta(t^0, x^0)\}$$

is differentiable on $N(a^0, r^0)$ with derivative

$$\nabla_{(a,r)} \tau^\delta(a, r) = \nabla_{(a,r)} \mathcal{L}(t(a,r), x(a,r), \mu(a,r), \nu(a,r), \xi(a,r), a, r).$$

Here $(t(a,r), x(a,r))$ denotes the local unique minimal solution of the optimization problem $(SP(a,r))$ with unique Lagrange multipliers $(\nu(a,r), \mu(a,r), \xi(a,r))$. Further the function $\phi \colon N(a^0, r^0) \to B_\delta(t^0, x^0) \times B_\delta(\mu^0, \nu^0, \xi^0)$ with

$$\phi(a, r) := (t(a,r), x(a,r), \mu(a,r), \nu(a,r), \xi(a,r))$$

is Lipschitz continuous on $N(a^0, r^0)$.

The result of Theorem 3.13 can also be found in Jittorntrum [131, Theorem 2]. In [78, Theorem 2.1] (see also the remark in [131, pp.128f]) Fiacco shows the differentiability of the minimal value function, too, but he assumes additionally non-degeneracy of the inequality constraints, i.e. $I^0 = \emptyset$ and $J^0 = \emptyset$. As already mentioned, in the case of non-degeneracy the index sets of the active and the inactive constraints remain unchanged under the assumptions of Theorem 3.13 (see also [79, Theorem 3.2.2c)]). Then we can examine problem $(SP(a, r))$ directly without switching to the modified problem: If we have for a minimal solution (t^0, x^0) of the reference problem $(SP(a^0, r^0))$

$$a_i^0 + t^0 r_i^0 - f_i(x^0) = 0 \quad \text{for} \quad i \in I^+,$$

then this holds also for the minimal solutions $(t(a,r), x(a,r))$ of $(SP(a,r))$ with (a, r) from a neighborhood of (a^0, r^0):

$$a_i + t(a,r)\, r_i - f_i(x(a,r)) = 0 \quad \text{for} \quad i \in I^+.$$

3.2 Sensitivity Results in Naturally Ordered Spaces

Therefore we can directly conclude from the local dependence of $t(a,r)$ on the parameters to the dependence of the points $f(x(a,r))$ on the parameters.

In the same way as in Corollary 3.10 we get the derivative of the local minimal value function in the point (a^0, r^0) with the help of the Lagrange multiplier μ^0 to the constraint $a^0 + t\,r^0 - f(x) \geq 0_m$:

Lemma 3.14. *Let the assumptions of Theorem 3.13 hold. Then the derivative of the local minimal value function in the point (a^0, r^0) is given by*

$$\nabla_{(a,r)} T^\delta(a^0, r^0) = \begin{pmatrix} -\mu^0 \\ -t^0 \mu^0 \end{pmatrix}.$$

Proof. Following the proof of Lemma 3.9 we get

$$\nabla_{(a,r)} T^\delta(a, r) = - \begin{pmatrix} \mu(a,r) \\ t(a,r)\,\mu(a,r) \end{pmatrix}$$

$$- \sum_{i=1}^{m} \nabla_{(a,r)} \mu_i(a,r)(a_i + t(a,r)\,r_i - f_i(x(a,r)))$$

$$- \sum_{j=1}^{p} \nabla_{(a,r)} \nu_j(a,r) g_j(x(a,r))$$

$$- \sum_{k=1}^{q} \nabla_{(a,r)} \xi_k(a,r) \underbrace{h_k(x(a,r))}_{=0}.$$

With that we obtain in the point (a^0, r^0)

$$\nabla_{(a,r)} T^\delta(a^0, r^0) = - \begin{pmatrix} \mu^0 \\ t^0 \mu^0 \end{pmatrix}$$

$$- \sum_{i \in I^+ \cup I^0} \nabla_{(a,r)} \mu_i(a^0, r^0) \underbrace{(a_i^0 + t^0 r_i^0 - f_i(x^0))}_{=0}$$

$$- \sum_{i \in I^-} \nabla_{(a,r)} \mu_i(a^0, r^0)(a_i^0 + t^0 r_i^0 - f_i(x^0))$$

$$- \sum_{j \in J^+ \cup J^0} \nabla_{(a,r)} \nu_j(a^0, r^0) \underbrace{g_j(x^0)}_{=0}$$

$$- \sum_{j \in J^-} \nabla_{(a,r)} \nu_j(a^0, r^0) g_j(x^0)$$

using $t(a^0, r^0) = t^0$, $x(a^0, r^0) = x^0$, and

$$(\mu(a^0, r^0), \nu(a^0, r^0), \xi(a^0, r^0)) = (\mu^0, \nu^0, \xi^0).$$

For $i \in I^-$ it is $a_i^0 + t^0 r_i^0 - f_i(x^0) > 0$ according to the definition of I^-. As the functions $a_i + t(a,r) r_i - f_i(x(a,r))$ are continuous in a and r there is a neighborhood $N(a^0, r^0)$ of (a^0, r^0) such that for all $(a, r) \in N(a^0, r^0)$ it holds

$$a_i + t(a,r) r_i - f_i(x(a,r)) > 0 \quad \text{for} \quad i \in I^-.$$

Then we conclude $\mu_i(a,r) = 0$ for all $(a,r) \in N(a^0, r^0)$ and hence $\nabla_{(a,r)} \mu_i(a^0, r^0) = 0_{2m}$ for $i \in I^-$. In the same way we can show $\nabla_{(a,r)} \nu_j(a^0, r^0) = 0_{2m}$ for $j \in J^-$ and we get

$$\nabla_{(a,r)} \tau^\delta(a^0, r^0) = \begin{pmatrix} -\mu^0 \\ -t^0 \mu^0 \end{pmatrix}.$$

□

For achieving better results in the approximation of the local behavior of the local minimal value function and with that of the values $t(a, r)$, information about higher order derivatives is useful. Having the second order derivative, i.e. the Hessian, we can give a second order approximation.

Theorem 3.15. *Let the assumptions of Theorem 3.13 hold. Additionally assume non-degeneracy of the inequality constraints, i.e. $I^0 = \emptyset$ and $J^0 = \emptyset$. The Hessian of the local minimal value function in the point (a^0, r^0) w.r.t. the parameter a is*

$$\nabla_a^2 \tau^\delta(a^0, r^0) = -\nabla_a \mu(a^0, r^0)$$

and w.r.t. the parameter r

$$\nabla_r^2 \tau^\delta(a^0, r^0) = t^0 \mu^0 (\mu^0)^\top - t^0 \nabla_r \mu(a^0, r^0).$$

Proof. According to Theorem 3.13 there is an open neighborhood $N(a^0, r^0)$ of (a^0, r^0), such that for $(a,r) \in N(a^0, r^0)$

$$\nabla_{(a,r)} \tau^\delta(a,r) = \nabla_{(a,r)} \mathcal{L}(t(a,r), x(a,r), \mu(a,r), \nu(a,r), \xi(a,r), a, r).$$

Using the same arguments as in the proof of Lemma 3.9 we conclude

$$\nabla_a \tau^\delta(a,r) = -\mu(a,r) - \sum_{i=1}^m \nabla_a \mu_i(a,r)(a_i + t(a,r) r_i - f_i(x(a,r)))$$

3.2 Sensitivity Results in Naturally Ordered Spaces

$$-\sum_{j=1}^{p} \nabla_a \nu_j(a,r) g_j(x(a,r)). \tag{3.9}$$

Besides, following again the steps of the proof of Lemma 3.9, we get for $(a,r) \in N(a^0, r^0)$ and $i \in I^-$

$$\nabla_a \mu_i(a,r) = 0_m \tag{3.10}$$

and for $j \in J^-$

$$\nabla_a \nu_j(a,r) = 0_m. \tag{3.11}$$

For $i \in I^+$ it is $\mu_i(a^0, r^0) > 0$. As the function $\mu(\cdot,\cdot)$ is continuous according to Theorem 3.13 there is a neighborhood $N^\mu(a^0, r^0) \subset N(a^0, r^0)$ of (a^0, r^0), such that for $(a,r) \in N^\mu(a^0, r^0)$ it holds

$$\mu_i(a,r) > 0$$

for $i \in I^+$ and thus

$$a_i + t(a,r) r_i - f_i(x(a,r)) = 0 \tag{3.12}$$

for all $(a,r) \in N^\mu(a^0, r^0)$, $i \in I^+$. Analogously there exists a neighborhood $N^\nu(a^0, r^0) \subset N(a^0, r^0)$ of (a^0, r^0) with $N^\nu(a^0, r^0) \subset N^\mu(a^0, r^0)$ and with

$$g_j(x(a,r)) = 0 \tag{3.13}$$

for all $(a,r) \in N^\nu(a^0, r^0)$, $j \in J^+$. As we consider the non-degenerated case, i.e. $I^+ \cup I^- = \{1,\ldots,m\}$ and $J^+ \cup J^- = \{1,\ldots,p\}$, we conclude from (3.9) using (3.10)-(3.13)

$$\nabla_a \tau^\delta(a,r) = -\mu(a,r).$$

In the case of non-degeneracy the function $\mu(\cdot,\cdot)$ is not only continuous but even continuously differentiable under the assumptions of the theorem (compare [79, Theorem 3.2.2.b)]), and we conclude

$$\nabla_a^2 \tau^\delta(a^0, r^0) = -\nabla_a \mu(a^0, r^0).$$

In the same way we can show

$$\nabla_r \tau^\delta(a,r) = -t(a,r) \mu(a,r).$$

Then we get for the Hessian of the local minimal value function with respect to r

$$\begin{aligned}\nabla_r^2 T^\delta(a^0, r^0) &= -t(a^0, r^0)\,\nabla_r \mu(a^0, r^0) - \mu(a^0, r^0)\,(\nabla_r t(a^0, r^0))^\top \\ &= -t^0\,\nabla_r \mu(a^0, r^0) - \mu^0\,(\nabla_r T^\delta(a^0, r^0))^\top \\ &= -t^0\,\nabla_r \mu(a^0, r^0) - \mu^0\,(-t(a^0, r^0)\,\mu(a^0, r^0))^\top \\ &= -t^0\,\nabla_r \mu(a^0, r^0) + t^0\,\mu^0(\mu^0)^\top.\end{aligned}$$

□

This result w.r.t. the parameter a, i.e. for the case that only the right hand side of the constraint $f_i(x) - t r_i \leq a_i$, $i = 1, \ldots, m$, is varied, can be found in [79, Corollary 3.4.4], too.

In Theorem 3.13 we also have the result that the function

$$\phi : N(a^0, r^0) \to B_\delta(t^0, x^0) \times B_\delta(\mu^0, \nu^0, \xi^0)$$

with

$$\phi(a, r) = (t(a, r), x(a, r), \mu(a, r), \nu(a, r), \xi(a, r))$$

is Lipschitz continuous on $N(a^0, r^0)$. Thereby $(t(a, r), x(a, r))$ denotes the local unique minimal solution of the optimization problem (SP(a, r)) with (a, r) from a neighborhood of (a^0, r^0) and $(\mu(a, r), \nu(a, r), \xi(a, r))$ are the correspondent unique Lagrange multipliers. We are interested in stronger results for getting an at least first-order local approximation of the function ϕ. These results will be needed for solving multiobjective bilevel optimization problems in the third part of this book. In the non-degenerated case we can apply a result by Fiacco ([79, Cor. 3.2.4]) to our scalarization approach:

Theorem 3.16. *Let the assumptions of Theorem 3.13 hold. Additionally assume non-degeneracy, i.e. $I^0 = \emptyset$ and $J^0 = \emptyset$. We consider the function $\phi : N(a^0, r^0) \to B_\delta(t^0, x^0) \times B_\delta(\mu^0, \nu^0, \xi^0)$ with*

$$\phi(a, r) = (t(a, r), x(a, r), \mu(a, r), \nu(a, r), \xi(a, r)).$$

Then we get the following first-order approximation for (a, r) in a neighborhood of (a^0, r^0):

$$\phi(a, r) = \phi(a^0, r^0) + M^{-1} N \begin{pmatrix} a - a^0 \\ r - r^0 \end{pmatrix} + o\left(\left\| \begin{pmatrix} a - a^0 \\ r - r^0 \end{pmatrix} \right\| \right)$$

with the matrix $M := [M_1 \mid M_2] \in \mathbb{R}^{(1+n+m+p+q) \times (1+n+m+p+q)}$ defined by

3.2 Sensitivity Results in Naturally Ordered Spaces

$$M_1 := \begin{bmatrix} 0 & 0_n^\top & -r_1 & \cdots & -r_m \\ 0_n & \nabla_x^2 \mathcal{L}(\phi(a^0, r^0), a^0, r^0) & \nabla_x f_1(x^0) & \cdots & \nabla_x f_m(x^0) \\ \mu_1^0 r_1 & -\mu_1^0 \nabla_x f_1(x^0)^\top & k_1^0 & 0\ldots 0 & 0 \\ \vdots & \vdots & \vdots & \ddots & \vdots \\ \mu_m^0 r_m & -\mu_m^0 \nabla_x f_m(x^0)^\top & 0 & 0\ldots 0 & k_m^0 \\ 0 & \nu_1^0 \nabla_x g_1(x^0)^\top & 0 & \cdots & 0 \\ \vdots & \vdots & \vdots & \ddots & \vdots \\ 0 & \nu_p^0 \nabla_x g_p(x^0)^\top & 0 & \cdots & 0 \\ 0 & \xi_1^0 \nabla_x h_1(x^0)^\top & 0 & \cdots & 0 \\ \vdots & \vdots & \vdots & \ddots & \vdots \\ 0 & \xi_p^0 \nabla_x h_q(x^0)^\top & 0 & \cdots & 0 \end{bmatrix}$$

and

$$M_2 := \begin{bmatrix} 0 & \cdots & 0 & 0 & \cdots & 0 \\ -\nabla_x g_1(x^0) & \cdots & -\nabla_x g_p(x^0) & -\nabla_x h_1(x^0) & \cdots & \nabla_x h_q(x^0) \\ 0_m & \cdots & 0_m & 0_m & \cdots & 0_m \\ g_1(x^0) & 0\ldots 0 & 0 & 0 & \cdots & 0 \\ \vdots & \ddots & \vdots & \vdots & \ddots & \vdots \\ 0 & 0\ldots 0 & g_p(x^0) & 0 & \cdots & 0 \\ 0_q & \cdots & 0_q & 0_q & \cdots & 0_q \end{bmatrix}$$

with $k^0 := a^0 + t^0 r^0 - f(x^0) \in \mathbb{R}^m$ and $N \in \mathbb{R}^{(1+n+m+p+q) \times (2m)}$ defined by

$$N := \left[\mathbf{0}_{2m \times (n+1)}, -\mu_1^0 \begin{pmatrix} e_1 \\ t^0 e_1 \end{pmatrix}, \ldots, -\mu_m^0 \begin{pmatrix} e_m \\ t^0 e_m \end{pmatrix}, \mathbf{0}_{2m \times (p+q)} \right]^\top.$$

Here, $\mathbf{0}_{2m \times (n+1)}$ and $\mathbf{0}_{2m \times (p+q)}$ denotes the matrix in $\mathbb{R}^{2m \times (n+1)}$ and $\mathbb{R}^{2m \times (p+q)}$, respectively, which has only the zero as entries, and e_i, $i = 1, \ldots, m$, denotes the ith unit vector in $\mathbb{R}^m$.

Based on this sensitivity information we can not only approximate the position of the point $f(x(a, r))$ (for (a, r) in a neighborhood of

$(a^0, r^0))$, but of the point $x(a,r)$ itself. We can use this for instance for calculating an advantageous starting point for a numerical method for solving the problem (SP(a,r)). The application of this approximation is discussed in Chap. 4 about the adaptive parameter control. There, an example is also studied on p. 112. A further application of this sensitivity theorem is considered in Chap. 7.

In Theorem 3.16 we have assumed non-degeneracy because otherwise the matrix M would not be invertible. However, we can still give a sensitivity result in the degenerate case. For that we assume that we restrict ourselves to a variation of the parameters (a,r) in one direction $v \in \mathbb{R}^{2m}$ only, i.e. we consider parameters (a,r) with

$$\begin{pmatrix} a \\ r \end{pmatrix} = \begin{pmatrix} a^0 \\ r^0 \end{pmatrix} + s\,v \quad \text{with } v = \begin{pmatrix} v^a \\ v^r \end{pmatrix} \in \mathbb{R}^{2m}$$

and $s \in \mathbb{R}$, $s \geq 0$, $v^a, v^r \in \mathbb{R}^m$, only. Then, a directional derivative of the function ϕ in direction v is sufficient. We get the directional derivative also in the degenerate case by applying a sensitivity result by Jittorntrum ([131, Theorem 3 and 4]). For that we consider the scalar optimization problem (SP(s))

$$\min t$$
subject to the constraints
$$(a^0 + s \cdot v^a) + t\,(r^0 + s \cdot v^r) - f(x) \geq_m 0_m,$$
$$g(x) \geq_p 0_p$$
$$x \in \mathbb{R}^n$$

depending on the parameter $s \in \mathbb{R}$, $s \geq 0$ only. We formulate this problem without equality constraints and for $v \in \mathbb{R}^{2m}$ constant. Let $(t(s), x(s))$ denote a minimal solution of (SP(s)) with Lagrange multipliers $(\mu(s), \nu(s))$. Then it is $(t(s), x(s), \mu(s), \nu(s)) = (t(a,r), x(a,r), \mu(a,r), \nu(a,r))$ for $(a,r) = (a^0 + s \cdot v^a, r^0 + s \cdot v^r)$.

Theorem 3.17. *Let the assumptions of Theorem 3.13 hold. Further consider the problem (SP(a,r)) without equality constraints. Let (t^0, x^0) be a minimal solution of (SP(a^0, r^0)) with Lagrange multipliers (μ^0, ν^0). We consider a variation of the parameters (a,r) restrained by*

$$\begin{pmatrix} a \\ r \end{pmatrix} = \begin{pmatrix} a^0 \\ r^0 \end{pmatrix} + s \cdot v$$

3.2 Sensitivity Results in Naturally Ordered Spaces

with $v = (v^a, v^r) \in \mathbb{R}^{2m}$ for $s \geq 0$.

Then there exists a unique solution $(\bar{t}, \bar{x}, \bar{\mu}, \bar{\nu})$ of the system of equalities and inequalities stated in (3.14) and in (3.15) and it is

$$\lim_{h \to 0^+} \begin{pmatrix} \frac{t(h)-t(0)}{h} \\ \frac{x(h)-x(0)}{h} \\ \frac{\mu(h)-\mu(0)}{h} \\ \frac{\nu(h)-\nu(0)}{h} \end{pmatrix} = \begin{pmatrix} \bar{t} \\ \bar{x} \\ \bar{\mu} \\ \bar{\nu} \end{pmatrix}$$

with $(t(s), x(s), \mu(s), \nu(s))$ the minimal solution and the correspondent Lagrange multipliers of the problem $(SP(a,r))$ with $(a,r) = (a^0, r^0) + s \cdot v$, $s \geq 0$.

The system of equalities and inequalities is given by

$$-\sum_{i=1}^{m} \bar{\mu}_i r_i^0 = \mu^{0\top} v^r,$$

$$\sum_{i=1}^{m} \mu_i^0 \nabla_x^2 f_i(x^0) \bar{x} - \sum_{j=1}^{p} \nu_j^0 \nabla_x^2 g_j(x^0) \bar{x}$$
$$+ \sum_{i=1}^{m} \bar{\mu}_i \nabla_x f_i(x^0) - \sum_{j=1}^{p} \bar{\nu}_j \nabla_x g_j(x^0) = 0_n,$$

$$r_i^0 \bar{t} - \nabla_x f_i(x^0)^\top \bar{x} = -v_i^a - t^0 v_i^r, \qquad \forall i \in I^+,$$
$$r_i^0 \bar{t} - \nabla_x f_i(x^0)^\top \bar{x} \geq -v_i^a - t^0 v_i^r, \qquad \forall i \in I^0,$$
$$\bar{\mu}_i \geq 0, \qquad \forall i \in I^0,$$
$$\bar{\mu}_i \left(r_i^0 \bar{t} - \nabla_x f_i(x^0)^\top \bar{x} + v_i^a + t^0 v_i^r \right) = 0, \qquad \forall i \in I^0,$$
$$\bar{\mu}_i = 0, \qquad \forall i \in I^-,$$

(3.14)

$$\nabla_x g_j(x^0)^\top \bar{x} = 0, \qquad \forall j \in J^+,$$
$$\nabla_x g_j(x^0)^\top \bar{x} \geq 0, \qquad \forall j \in J^0,$$
$$\bar{\nu}_j \geq 0, \qquad \forall j \in J^0, \qquad (3.15)$$
$$\bar{\nu}_j \left(\nabla_x g_j(x^0)^\top \bar{x} \right) = 0, \qquad \forall j \in J^0,$$
$$\bar{\nu}_j = 0, \qquad \forall j \in J^-.$$

We have formulated this theorem for the case without equality constraints, but as the degenerated case is included equality constraints can be handled as well. We continue the examination of sensitivity results for scalarizations to naturally ordered multiobjective optimization problems in the following section.

3.3 Sensitivity Results for the ε-Constraint Problem

In Sect. 2.5 we have already discussed, that many scalarization approaches like the ε-constraint problem can be seen as a special case of the Pascoletti-Serafini scalarization. Thus, we can apply the results of the preceding sections to these scalarizations, too. We demonstrate this on the ε-constraint problem for the case that the ordering induced by K equals the natural ordering. Then we gain results on the connection between the minimal value of the ε-constraint problem $(P_k(\varepsilon))$ and the parameters ε_i, $i \in \{1, \ldots, m\} \setminus \{k\}$. For convenience we recall the ε-constraint scalarization $(P_k(\varepsilon))$

$$\min f_k(x)$$
$$\text{subject to the constraints}$$
$$f_i(x) \leq \varepsilon_i, \quad i \in \{1, \ldots, m\} \setminus \{k\},$$
$$g_j(x) \geq 0, \quad j = 1, \ldots, p, \quad (3.16)$$
$$h_l(x) = 0, \quad l = 1, \ldots, q,$$
$$x \in \mathbb{R}^n.$$

We assume in this section $S = \mathbb{R}^n$. We apply Theorem 3.13 and Lemma 3.14:

Theorem 3.18. *Let the Assumption 3.12 hold. Let x^0 be a local minimal solution of the reference problem $(P_k(\varepsilon^0))$ with Lagrange multipliers $\mu_i^0 \in \mathbb{R}_+$ for $i \in \{1, \ldots, m\} \setminus \{k\}$, $\nu^0 \in \mathbb{R}_+^p$, and $\xi^0 \in \mathbb{R}^q$. Let there exist a $\gamma > 0$ so that the functions f, g, and h are twice continuously differentiable on an open neighborhood of the closed ball $B_\gamma(x^0)$. Let the index sets $I^+ \cup I^0 \cup I^- = \{1, \ldots, m\} \setminus \{k\}$ and $J^+ \cup J^0 \cup J^- = \{1, \ldots, p\}$ be defined by*

$$I^+ := \{i \in \{1, \ldots, m\} \setminus \{k\} \mid f_i(x^0) = \varepsilon_i^0, \ \mu_i^0 > 0\},$$
$$I^0 := \{i \in \{1, \ldots, m\} \setminus \{k\} \mid f_i(x^0) = \varepsilon_i^0, \ \mu_i^0 = 0\}, \quad (3.17)$$
$$I^- := \{i \in \{1, \ldots, m\} \setminus \{k\} \mid f_i(x^0) < \varepsilon_i^0, \ \mu_i^0 = 0\}$$

and (3.8).
Further assume the following:

3.3 Sensitivity Results for the ε-Constraint Problem 95

a) The gradients of the (in the point x^0) active constraints, i. e. the vectors $\nabla_x f_i(x^0)$ for $i \in I^+ \cup I^0$, $\nabla_x g_j(x^0)$ for $j \in J^+ \cup J^0$, and $\nabla_x h_l(x^0)$ for $l \in \{1,\ldots,q\}$ are linearly independent.

b) There exists a constant $\alpha > 0$ such that

$$x^\top \nabla_x^2 \hat{\mathcal{L}}(x^0, \mu^0, \nu^0, \xi^0, \varepsilon^0) \, x \geq \alpha \|x\|^2$$

for all

$$x \in \{x \in \mathbb{R}^n \mid \nabla_x f_i(x^0) x = 0, \quad \forall i \in I^+, \, \nabla_x g_j(x^0) x = 0, \, \forall j \in J^+,$$
$$\nabla_x h_l(x^0) x = 0, \quad \forall l = 1,\ldots,q\} =: W. \tag{3.18}$$

with $\nabla_x^2 \hat{\mathcal{L}}$ the Hessian of the Lagrange function in the point x^0 of the problem $(P_k(\varepsilon^0))$.

Then x^0 is a local unique minimal solution of $(P_k(\varepsilon^0))$ with unique Lagrange multipliers and there is a $\delta > 0$ and a neighborhood $N(\varepsilon^0)$ of ε^0 so that the local minimal value function $\hat{\tau}^\delta : \mathbb{R}^{m-1} \to \overline{\mathbb{R}}$,

$$\hat{\tau}^\delta(\varepsilon) := \inf\{f_k(x) \mid f_i(x) \leq \varepsilon_i, \, i \in \{1,\ldots,m\} \setminus \{k\},$$
$$g_j(x) \geq 0, \, j = 1,\ldots,p,$$
$$h_l(x) = 0, \, l = 1,\ldots,q, \, x \in B_\delta(x^0)\}$$

is differentiable on $N(\varepsilon^0)$ and

$$\frac{\partial \hat{\tau}^\delta(\varepsilon^0)}{\partial \varepsilon_i} = -\mu_i^0.$$

Proof. According to Lemma 2.27 parameters (a^0, r^0) exist so that the problem $(P_k(\varepsilon^0))$ is equivalent to problem $(\text{SP}(a^0, r^0))$. We show that the assumptions of Theorem 3.13 are satisfied for the problem $(\text{SP}(a^0, r^0))$. It remains to show that assumption b) of Theorem 3.13 is implicated by assumption b) of this theorem. Let $\mathcal{L}$ denote the Lagrange function of the problem $(\text{SP}(a^0, r^0))$ with a^0, r^0 defined by ε^0 as in (2.25). It is

$$\nabla^2_{(t,x)} \mathcal{L}(f_k(x^0), x^0, \mu^0, \nu^0, \xi^0, a^0, r^0) = \begin{pmatrix} 0 & 0 \\ 0 & W \end{pmatrix}$$

with $W := \nabla_x^2 \mathcal{L}(f_k(x^0), x^0, \mu^0, \nu^0, \xi^0, a^0, r^0)$ and so it is sufficient to show that there exists a $\beta > 0$ with

$$x^\top W x \geq \beta \left\| \begin{pmatrix} t \\ x \end{pmatrix} \right\|^2$$

for all

$$(t, x) \in \{(t, x) \in \mathbb{R} \times \mathbb{R}^n \mid x \in \mathcal{W}, \nabla_x f_k(x^0)^\top x = t\}.$$

For the definition of $\mathcal{W}$ see (3.18). Using $\nabla_x f_k(x^0)^\top x = t$ it is sufficient to show

$$x^\top W x \geq \beta \left\| \begin{pmatrix} \nabla_x f_k(x^0) x \\ x \end{pmatrix} \right\|^2$$

for all $x \in \mathcal{W}$. Since in $\mathbb{R}^n$ and $\mathbb{R}^{n+1}$ respectively all norms are equivalent there exist positive constants $M^l, M^u \in \mathbb{R}$ and $\tilde{M}^l, \tilde{M}^u \in \mathbb{R}$ with

$$M^l \|x\|_2 \leq \|x\| \leq M^u \|x\|_2$$

and

$$\tilde{M}^l \left\| \begin{pmatrix} \nabla_x f_k(x^0)^\top x \\ x \end{pmatrix} \right\|_2 \leq \left\| \begin{pmatrix} \nabla_x f_k(x^0)^\top x \\ x \end{pmatrix} \right\| \leq \tilde{M}^u \left\| \begin{pmatrix} \nabla_x f_k(x^0)^\top x \\ x \end{pmatrix} \right\|_2.$$

We set

$$\beta := \frac{\alpha (M^l)^2}{(\tilde{M}^u)^2 (\|\nabla_x f_k(x^0)\|_2^2 + 1)} > 0.$$

It is $W = \nabla_x^2 \hat{\mathcal{L}}(x^0, \mu^0, \nu^0, \xi^0, \varepsilon^0)$ and thus we can conclude from assumption b) of this theorem

$$x^\top W x \geq \alpha \|x\|^2$$
$$\geq \alpha (M^l)^2 \|x\|_2^2$$
$$= \beta (\tilde{M}^u)^2 (\|\nabla_x f_k(x^0)\|_2^2 + 1) \|x\|_2^2$$
$$\geq \beta (\tilde{M}^u)^2 \left(|\nabla_x f_k(x^0)^\top x|^2 + \|x\|_2^2 \right)$$
$$= \beta (\tilde{M}^u)^2 \left\| \begin{pmatrix} \nabla_x f_k(x^0)^\top x \\ x \end{pmatrix} \right\|_2^2$$
$$\geq \beta \left\| \begin{pmatrix} \nabla_x f_k(x^0)^\top x \\ x \end{pmatrix} \right\|^2$$

for all $x \in \mathcal{W}$. Then the assumptions of Theorem 3.13 are satisfied. With the help of Lemma 3.14 we can calculate the partial derivatives of $\hat{\tau}^\delta$ w.r.t. ε_i. □

3.3 Sensitivity Results for the ε-Constraint Problem

We get this specialized result by directly applying a sensitivity theorem by Luenberger ([157, p.236]), too. Polak uses in [182] also this sensitivity theorem by Luenberger and applies it to the ε-constraint problem $(P_k(\varepsilon))$ with $k = m$. He uses the derivative of the local minimal value function given by the Lagrange multipliers for an approximation of this function based on a cubic Hermite interpolation. He estimates the interpolation error and tries to limit this error by an appropriate parameter choice. Polak calls the local minimal value function a sensitivity function.

Chankong and Haimes use this sensitivity result ([28, p.58]) for an interpretation of the Lagrange multipliers as trade-off information between the competing objective functions ([28, p.160 and Sect. 7.4.1]). They also consider the linear case ([28, Theorem 4.31]) and finally develop an interactive method, the so-called surrogate worth trade-off method ([28, Chap. 8], [99]). This method is based on the ε-constraint method with an equidistant parameter choice and an interrogation of the decision maker regarding the function values together with the trade-off information.

In the dissertation of Heseler, [109], also scalarization approaches to multiobjective optimization problems are considered as parameter dependent scalar problems. Then warm-start strategies for interior-point methods are applied. There stability considerations ([109, Theorem 2.2.2]) play an important role, too. Finally bicriteria convex quadratic optimization problems are solved by using the weighted sum method.

The sensitivity result of Theorem 3.18 can be used for approximating the position of the weakly efficient points in dependence on the parameter ε. Let x^0 be a minimal solution of the problem $(P_k(\varepsilon^0))$ with Lagrange multipliers μ_i^0 to the constraints $f_i(x) \leq \varepsilon_i^0$ for $i = \{1, \ldots, m\} \setminus \{k\}$. Then we get for the points $f_k(x(\varepsilon))$ with $x(\varepsilon)$ the minimal solution of $(P_k(\varepsilon))$ for ε in a neighborhood of ε^0:

$$f_k(x(\varepsilon)) \approx f_k(x^0) - \sum_{\substack{i=1 \\ i \neq k}}^{m} \mu_i^0 (\varepsilon_i - \varepsilon_i^0).$$

For the (in the point x^0) active constraints we have

$$f_i(x(\varepsilon)) = \varepsilon_i$$

because active constraints remain active.

The assumptions of Theorem 3.18 are not too restrictive. In many applications it turns out that the efficient set is smooth, see for instance [16, 67]. The efficient set corresponds directly to the solutions and minimal values of the ε-constraint scalarization for varying parameters. Thus differentiability of the minimal-value function w.r.t. the parameters can be presumed in many cases.

The sensitivity results of this chapter can be used to control the choice of the parameters a and ε, respectively, adaptively as described in the following chapter.

Part II

Numerical Methods and Results

4
Adaptive Parameter Control

In this chapter we use the preceding results for developing an algorithm for adaptively controlling the choice of the parameters in several scalarization approaches. The aim is an approximation of the efficient set of the multiobjective optimization problem with a high quality. The quality can be measured with different criteria, which we discuss first. This leads us to the aim of equidistant approximation points.

For reaching this aim we mainly use the scalarization approach of Pascoletti and Serafini. This scalarization is parameter dependent and we develop a procedure how these parameters can be chosen adaptively such that the distances between the found approximation points of the efficient set are controlled. For this adaptive parameter choice we apply the sensitivity results of Chap. 3. Because many other scalarizations can be considered as a special case of the Pascoletti-Serafini problem, as we have seen in Sect. 2.5, we can apply our results for the adaptive parameter control to other scalarizations as the ε-constraint or the normal boundary intersection problem, too.

4.1 Quality Criteria for Approximations

The importance of a representative approximation of the whole efficient set is often pointed out, see for instance [10, 13, 39, 77, 81, 82, 201]. In many works which present a numerical method for solving multiobjective optimization problems it is the aim to generate nearly equidistant approximations (for instance in [40, 44, 138, 163, 164]) to obtain a representative but concise approximation and thus a high quality of the approximation. For being able to measure such a quality we discuss several quality criteria in the following. First, we define what we mean by an approximation (compare [101, p.5]).

Definition 4.1. A finite set $A \subset f(\Omega)$ is called an approximation of the efficient set $\mathcal{E}(f(\Omega), K)$ of the multiobjective optimization problem (MOP), if for all approximation points $y^1, y^2 \in A$, $y^1 \neq y^2$ it holds

$$y^1 \notin y^2 + K \text{ and } y^2 \notin y^1 + K, \tag{4.1}$$

i. e. all points in A are non-dominated to each other w. r. t. the ordering cone K.

Following this definition we call a finite set A an approximation of the weakly efficient set $\mathcal{E}_w(f(\Omega), K)$ of the multiobjective optimization problem (MOP) if A is an approximation of $\mathcal{E}(f(\Omega), \text{int}(K) \cup \{0_m\})$. Then the points in A are non-dominated to each other w. r. t. the interior of the cone K.

Here, we generally consider approximations of the efficient set, i. e. of the image of the set of K-minimal points, but the definition as well as the following quality criteria can be transferred to approximations of the set of K-minimal points, too. We concentrate our considerations to the efficient set as a decision maker, who is asked to select his or her subjectively preferred solution, usually makes his decision based on a comparison of the function values in the image and not in the preimage set.

Besides, the dimension n of the preimage space is generally distinctly higher than the dimension m of the image space. For example in Chap. 6 we discuss an application problem in medical engineering with $m = 2$ but $n = 400$. An approximation of the K-minimal solutions in $\mathbb{R}^n$ is then not visualizable and besides not interpretable by the decision maker.

A third reason is, that often two K-minimal solutions $x, y \in \Omega$ have the same objective value $f(x) = f(y)$. Then, w. r. t. the objective functions, these two points are no longer distinguishable, but the set which has to be approximated (in the preimage space) is enlarged. A discussion of these reasons is also done by Benson and Sayin in [10]. However there also exist problems which demand an approximation of the K-minimal points and not of the efficient set with a controlled quality. In Chap. 7 we study multiobjective bilevel optimization problems where this occurs. Nevertheless, generally we concentrate on the efficient set. We generate approximations of the weakly efficient set and choose from that set due to $\mathcal{E}(f(\Omega),K) \subset \mathcal{E}_w(f(\Omega),K)$ an approximation of $\mathcal{E}(f(\Omega),K)$. Here, we also call the set

$$\{x \in \Omega \mid f(x) \in A\}$$

an approximation if A is an approximation.

According to the definition, approximations do not need to consist of efficient points of $f(\Omega)$. Non-efficient approximation points are often generated by methods based on evolutionary algorithms as presented in [31, 87, 228, 246], and others. Then the distance between the approximation points and the efficient set is proposed as a quality criteria, compare [32, 93, 141],[228, p.6-15] and [246, p.46].

In the literature related to numerical methods based on evolutionary algorithms for solving multiobjective optimization problems various other criteria for measuring and comparing the approximation qualities are given, for instance in [43, 93, 141, 228, 246]. However, due to the fact that in numerical methods based on scalarization approaches the approximation points are determined by solving these scalar problems, the points are generally at least weakly K-minimal (ignoring small inaccuracies caused by the used numerical solvers). For the Pascoletti-Serafini problem this is the case according to Theorem 2.1,c). Then we have $A \subset \mathcal{E}_w(f(\Omega),K)$ and thus measuring the distance of A to the efficient set is not an interesting criteria. Note, that the points which are generated by the Pascoletti-Serafini scalarization satisfy directly the property (4.1) w.r.t. $\operatorname{int}(K) \cup \{0_m\}$.

We restrain ourselves on criteria which are meaningful in our case and present three quality criteria defined by Sayin in [191] called coverage error, uniformity and cardinality. The coverage error is used for measuring whether the approximation is representative.

Definition 4.2. Let $\varepsilon > 0$. An approximation A of the efficient set $\mathcal{E}(f(\Omega),K)$ is called d_ε-representation of $\mathcal{E}(f(\Omega),K)$, if for all $y \in \mathcal{E}(f(\Omega),K)$ there exists a point $\bar{y} \in A$ with $\|y - \bar{y}\| \leq \varepsilon$. The smallest ε for which A is a d_ε-representation of $\mathcal{E}(f(\Omega),K)$ is called coverage error of A.

The coverage error can be calculated by

$$\varepsilon = \max_{y \in \mathcal{E}(f(\Omega),K)} \min_{\bar{y} \in A} \|y - \bar{y}\|. \tag{4.2}$$

Here the knowledge of the efficient set $\mathcal{E}(f(\Omega),K)$ is needed. If this is the case then (4.2) is, for a set $A = \{y^1, \ldots, y^N\}$, equivalent to

$$\varepsilon = \max \delta$$

subject to the constraints

$$\delta \leq \|y - y^i\|, \quad i = 1, \ldots, N, \qquad (4.3)$$
$$y \in \mathcal{E}(f(\Omega), K),$$
$$\delta \in \mathbb{R}.$$

Generally the efficient set is not known but then it is important still to be able to calculate or at least to approximate the coverage error. For doing this we assume that we have already assured by the applied method that the whole efficient set is covered by the approximation and that no larger parts are neglected. Then the criteria of a small coverage error corresponds to the target, that the distance between nearby approximation points is as small as possible.

We have to take into account that the efficient set is not necessarily connected and that gaps can exist as it is the case in Example 4.4. Conditions under which the efficient set is connected are given in [105, 113, 156, 172]. Thereby a set A is connected (in the topological sense) if no open sets O_1, O_2 exist such that $A \subset O_1 \cup O_2$, $A \cap O_1 \neq \emptyset$, $A \cap O_2 \neq \emptyset$, and $A \cap O_1 \cap O_2 = \emptyset$ ([60, p.69], [190, p.66]). A large distance between consecutive approximation points in non-connected parts of the efficient set can thus not be avoided if this is due to the size of the gap. This should not affect the coverage error. This fact is also mentioned by Collette in [32, p.780].

If the efficient set is connected we can calculate an approximation of the coverage error, called modified coverage error, without knowing the efficient set explicitly by

$$\bar{\varepsilon} := \frac{1}{2} \max_{j \in \{1, \ldots, N\}} \max_{y \in \mathcal{N}(y^j)} \|y^j - y\|. \qquad (4.4)$$

Here, $\{y^1, \ldots, y^N\}$ denotes the approximation A and $\mathcal{N}(y^j)$ denotes the subset of points of A which are in a neighborhood of the point y^j. In the following we define which points are included in the set $\mathcal{N}(y^j)$. We always assume $N > 1$.

For the case $m = 2$ we can (without loss of generality) order the approximation points by

$$y_1^1 \leq y_1^2 \leq \ldots \leq y_1^N.$$

Then let $\mathcal{N}(y^j) := \{y^{j-1}, y^{j+1}\}$ for $j \in \{2, \ldots, N-1\}$, and $\mathcal{N}(y^1) := \{y^2\}$ and $\mathcal{N}(y^N) := \{y^{N-1}\}$, respectively. For $m \geq 3$ we choose the

$2(m-1)$ approximation points with the smallest distance to the point y^j as neighbors for the set $\mathcal{N}(y^j)$. If the efficient set is not connected this has to be taken into account for the determination of the sets $\mathcal{N}(\cdot)$.

The smaller the coverage error the better any point of the efficient set is represented by the approximation. The importance of a coverage of the efficient set without gaps is often emphasized in applications. For instance in [143, p.231] the definition of a ϱ-cover as a set of points such that any point of this set is within the distance of ϱ to at least one other point of the set is given (compare also [83, p. 23]).

A further criteria for a high approximation quality is the uniformity of the distribution of the approximation points. If the points are too dense these additional points deliver no new information but demand a high computational effort. In the sense of uniformity an ideal approximation is thus an equidistant approximation. The aim of equidistancy is also proposed in [109, p.59]. As a definition for the uniformity level δ the following is given:

Definition 4.3. Let A be a d_ε-representation of $\mathcal{E}(f(\Omega), K)$ with coverage error ε. Then A is a δ-uniform d_ε-representation if

$$\min_{\substack{x,y \in A \\ x \neq y}} \|x - y\| \geq \delta.$$

The largest δ for which A is a δ-uniform d_ε-representation is called uniformity level of the approximation A.

The uniformity level δ can easily be calculated by

$$\delta = \min_{\substack{x,y \in A \\ x \neq y}} \|x - y\|.$$

A high value for δ corresponds to a high uniformity.

The third criteria, the cardinality of the approximation, is measured by the number of different points in A. Thus the maximal cardinality is N. The aim is to give a representative approximation with as few points as possible. The following example clarifies the three discussed quality criteria ([188, p.67]).

Example 4.4. We consider the approximation of the efficient set of the bicriteria optimization problem with $K = \mathbb{R}^2_+$ shown in Fig. 4.1 consisting of the three points y^1, y^2, y^3. Thus the cardinality is 3. We use the Euclidean norm. The uniformity level δ is the smallest distance between two approximation points and is thus

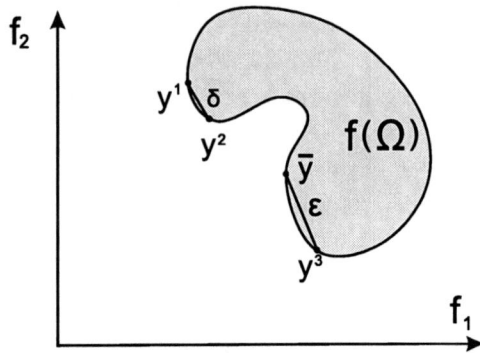

Fig. 4.1. Approximation of Example 4.4.

$$\delta = \|y^1 - y^2\|_2.$$

The coverage error is the distance between the worst represented efficient point and the closest approximation point and thus

$$\varepsilon = \|y^3 - \bar{y}\|_2.$$

Obviously the quality criteria are conflicting and the aim of a high quality of an approximation is thus by itself a multiobjective optimization problem. For instance a small coverage error corresponds to a high cardinality. As a consequence not all approximations can be compared and ordered as the following example demonstrates.

Example 4.5. We consider the approximations

$$A^1 = \{y^2, y^4, y^6, y^8\} \quad \text{and}$$
$$A^2 = \{y^3, y^7\}.$$

of the efficient set of the bicriteria optimization problem with $K = \mathbb{R}^2_+$ shown in Fig. 4.2. We assume that it holds for a $\beta > 0$

$$\|y^{i+1} - y^i\|_2 = \frac{\beta}{2} \text{ for } i = 1, \ldots, 8.$$

Then $\varepsilon^1 = \frac{\beta}{2}$ is the modified coverage error of A^1 and $\varepsilon^2 = \beta$ is the modified coverage error of A^2. The uniformity level of the approximation A^1 is given by $\delta^1 = \beta$ and of A^2 by $\delta^2 = 2\beta$.

Both approximations are optimal w. r. t. the third criteria, the cardinality, i. e. there is no other approximation with coverage error ε^1 and

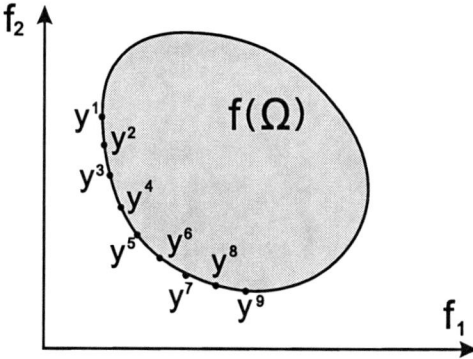

Fig. 4.2. Approximations of Example 4.5.

uniformity level δ^1 and ε^2 and δ^2, respectively, with a smaller number of points. Here no decision which approximation has a higher quality can be made.

Thus, a decision maker cannot just ask for the general best approximation but he can only ask for instance for a best approximation w. r. t. the coverage error for a given cardinality. We therefore concentrate on the aim of an equidistant approximation with the possibility that the decision maker chooses the distance $\alpha > 0$ between the points. That is, neighbored points should have a distance of α to each other. Thus the modified coverage error (4.4) is aimed to be as small as possible, but a value of $\alpha/2$ is sufficient. At the same time the uniformity level should have a value of α to avoid a too high density of the points and to reach the aim of a distance of α between the approximation points. The cardinality is then implicitly defined by the distance α.

In the following sections in this chapter we propose a method for reaching the aim of equidistant approximation points by an adaptive parameter choice in the used scalarization approaches. We use the sensitivity results of Chap. 3 for developing that adaptive parameter control. We start with the scalarization by Pascoletti and Serafini and in the consecutive sections we apply the gained results to further scalarization problems. First we discuss the bicriteria case.

4.2 Adaptive Parameter Control in the Bicriteria Case

In this section we consider the bicriteria case only, i. e. we consider multiobjective optimization problems (MOP) with two objective functions

4 Adaptive Parameter Control

$(m = 2)$:

$$\min \begin{pmatrix} f_1(x) \\ f_2(x) \end{pmatrix}$$

subject to the constraint

$$x \in \Omega.$$

The aim is an approximation of the efficient set of this problem for which we solve the already discussed scalar problems (SP(a,r))

$$\min t$$

subject to the constraints

$$a + tr - f(x) \in K,$$
$$t \in \mathbb{R}, \ x \in \Omega.$$

By a variation of the parameter a several approximation points can be found. As we do not want arbitrarily distributed approximation points we develop a procedure for controlling the choice of the parameter a. This procedure is an adaptive method based on the gained sensitivity information and the only interaction by the d.m. needed is the pre-defining of the desired distance α.

In the following we assume the parameter r to be chosen constant with $r \in K \setminus \{0_2\}$. We choose the parameter a from the hyperplane

$$H = \{y \in \mathbb{R}^2 \mid b^\top y = \beta\} = \{y \in \mathbb{R}^2 \mid b_1 y_1 + b_2 y_2 = \beta\},$$

which is a line here, with $b \in \mathbb{R}^2$, $\beta \in \{0, 1\}$, and $b^\top r \neq 0$, as discussed in Sect. 2.3.1. According to Theorem 2.17 it is sufficient to consider only parameters $a \in H$ with $a \in H^a \subset H$ for

$$H^a = \{y \in \mathbb{R}^2 \mid y = \lambda \bar{a}^1 + (1 - \lambda) \bar{a}^2, \ \lambda \in [0, 1]\}.$$

Without loss of generality we can assume $\bar{a}_1^1 < \bar{a}_1^2$. We now want to determine parameters a^0, a^1, a^2, ... adaptively (starting with $a^0 = \bar{a}^1$) such that the related approximation points $f(x^i)$, $i = 0, 1, 2, \ldots$, gained by solving (SP(a^i, r)) for $i = 0, 1, 2, \ldots$, have the equal distance $\alpha > 0$. We assume further $a_1^0 \leq a_1^1 \leq a_1^2 \leq \ldots$. For reaching this aim we use the sensitivity results for the problems (SP(a,r)) and ($\overline{\text{SP}}(a,r)$) of Chap. 3. For that we need the following assumptions:

Assumption 4.6. Let the Assumption 3.4 hold. Moreover, to any parameters a and r for which we consider the optimization problem

4.2 Parameter Control in the Bicriteria Case

(SP(a,r)) and ($\overline{\text{SP}}(a,r)$) respectively let there exist a minimal solution $(\bar{t}, \bar{x})$ with Lagrange multipliers $(\bar{\mu}, \bar{\nu}, \bar{\xi}) \in \mathbb{R}^m \times C^* \times \mathbb{R}^q$ and let the assumptions of Theorem 3.7 be satisfied in $(\bar{t}, \bar{x})$. Further let $S = \mathbb{R}^n$.

We assume we have already solved the problem (SP(a^0, r)) for a parameter $a^0 \in H^a$ with a minimal solution (t^0, x^0) and Lagrange multipliers (μ^0, ν^0, ξ^0). In the following paragraphs we examine how the next parameter a^1 should be chosen.

First we consider the case that the constraint $a^0 + tr - f(x) \in K$ is active in the point (t^0, x^0), i.e.

$$a^0 + t^0 r - f(x^0) = 0_2. \tag{4.5}$$

Later we will see that the case $a^0 + t^0 r - f(x^0) \neq 0_2$ can easily be transferred to this case. Due to (4.5) the point (t^0, x^0) is also a minimal solution of the modified problem ($\overline{\text{SP}}(a^0, r)$)

$$\min t$$

subject to the constraints
$$a^0 + tr - f(x) = 0_2$$
$$t \in \mathbb{R},\ x \in \Omega$$

with Lagrange multipliers (μ^0, ν^0, ξ^0) according to Lemma 2.26.

We assume, as a consequence of Theorem 3.7, that the derivative $\nabla_a \bar{\tau}^\delta(a^0, r)$ of the local minimal value function in the point (a^0, r) is known. We use this derivative information for a local approximation of the local minimal value function to the optimization problem ($\overline{\text{SP}}(a, r)$). We presume that we can do such a first order Taylor approximation and then we get by using $\bar{\tau}^\delta(a^0, r) = \bar{t}(a^0, r) = t^0$ (with $(\bar{t}(a, r), \bar{x}(a, r))$ the local minimal solution of ($\overline{\text{SP}}(a, r)$))

$$\bar{t}(a, r) \approx t^0 + \nabla_a \bar{\tau}^\delta(a^0, r)^\top (a - a^0). \tag{4.6}$$

As we consider the scalar problem ($\overline{\text{SP}}(a, r)$) we have the equation

$$f(\bar{x}(a, r)) = a + \bar{t}(a, r)\, r$$

and we obtain together with $f(x^0) = a^0 + t^0 r$

$$f(\bar{x}(a, r)) \approx a^0 + (a - a^0) + \left(t^0 + \nabla_a \bar{\tau}^\delta(a^0, r)^\top (a - a^0)\right)\, r$$
$$= f(x^0) + (a - a^0) + \left(\nabla_a \bar{\tau}^\delta(a^0, r)^\top (a - a^0)\right)\, r.$$

4 Adaptive Parameter Control

We use the point $(\bar{t}(a,r), \bar{x}(a,r))$ as an approximation of the minimal solution $(t(a,r), x(a,r))$ of $(SP(a,r))$ (we have at least $t(a,r) \leq \bar{t}(a,r)$ and $(\bar{t}(a,r), \bar{x}(a,r))$ is feasible for $(SP(a,r))$). Then we obtain the following approximation for the weakly K-efficient points of the multiobjective optimization problem dependent on the parameter a (for a in a neighborhood of a^0):

$$f(x(a,r)) \approx f(x^0) + (a - a^0) + \left(\nabla_a \bar{\tau}^\delta(a^0, r)^\top (a - a^0)\right) r. \quad (4.7)$$

The aim is to determine the next parameter a^1 so that for the given distance $\alpha > 0$ we obtain

$$\|f(x(a^0, r)) - f(x(a^1, r))\| = \alpha \quad (4.8)$$

with $f(x(a^0, r)) = f(x^0)$. That is, the new approximation point shall have a distance of α to the former.

Further let $a^1 \in H$. For that we choose a direction $v \in \mathbb{R}^2 \setminus \{0_2\}$ with $a^0 + sv \in H$ for $s \in \mathbb{R}$, i.e. let $b^\top v = 0$. Besides, as we also have presumed $a_1^0 \leq a_1^1$, we demand $s \geq 0$ and $v_1 \geq 0$. This is satisfied for instance for $v := \bar{a}^2 - \bar{a}^1$. We now search for a scalar $s^1 \in \mathbb{R}$ such that $a^1 := a^0 + s^1 v$ satisfies (4.8). As $a^1 - a^0 = s^1 v$ we get, based on the approximation (4.7),

$$\begin{aligned}
\alpha &= \left\|f(x(a^0, r)) - f(x(a^1, r))\right\| \\
&\approx \left\|f(x^0) - \left(f(x^0) + s^1 v + s^1 \left(\nabla_a \bar{\tau}^\delta(a^0, r)^\top v\right) r\right)\right\| \\
&= |s^1| \left\|v + \left(\nabla_a \bar{\tau}^\delta(a^0, r)^\top v\right) r\right\|.
\end{aligned}$$

Thus, for satisfying (4.8), we choose

$$s^1 := \frac{\alpha}{\|v + \left(\nabla_a \bar{\tau}^\delta(a^0, r)^\top v\right) r\|} > 0. \quad (4.9)$$

We define $a^1 := a^0 + s^1 v$. Solving $(SP(a^1, r))$ results in a weakly K-minimal point x^1. Dependent on the quality of the approximation in (4.7) we have

$$\|f(x^0) - f(x^1)\| \approx \alpha.$$

For the calculation of s^1 the derivative $\nabla_a \bar{\tau}^\delta(a^0, r)$ has to be known. In the case of the ordering cone $C = \mathbb{R}_+^p$ we have according to Corollary 3.10

4.2 Parameter Control in the Bicriteria Case

$$\nabla_a \bar{\tau}^\delta(a^0, r) = -\mu^0.$$

For an arbitrary cone C the expression

$$\nabla_a \bar{\tau}^\delta(a^0, r) = -\mu^0 - \nabla_a \nu(a^0, r)^\top g(x^0)$$

cannot be simplified generally. Then, the derivative of the function $\nu(\cdot, \cdot)$ in the point (a^0, r) w.r.t. the parameter a can be approximated numerically for instance by numerical differentiation.

The approximation of the local minimal value function may be improved by using higher order derivatives. For example for $K = \mathbb{R}^2_+$, $C = \mathbb{R}^p_+$ and in the case of non-degeneracy of the constraints the second order derivative is given by

$$\nabla_a^2 \bar{\tau}^\delta(a^0, r) = -\nabla_a \mu(a^0, r)$$

according to Theorem 3.15. Here, the derivative of $\mu(\cdot, \cdot)$ in the point (a^0, r) w.r.t. a can again be approximated by numerical differentiation. We obtain

$$\bar{t}(a, r) \approx t^0 + \nabla_a \bar{\tau}^\delta(a^0, r)^\top (a - a^0) + \frac{1}{2}(a - a^0)^\top \nabla_a^2 \bar{\tau}^\delta(a^0, r)(a - a^0)$$

and thus for $a^1 = a^0 + s^1 v$

$$\alpha = \|f(x(a^0, r)) - f(x(a^1, r))\|$$
$$\approx \left\| f(x^0) - \left(f(x^0) + s^1 v + s^1 \left(\nabla_a \bar{\tau}^\delta(a^0, r)^\top v \right) r \right. \right.$$
$$\left. \left. + \tfrac{1}{2}(s^1)^2 \left(v^\top \nabla_a^2 \bar{\tau}^\delta(a^0, r) v \right) r \right) \right\|$$
$$= \left\| s^1 v + s^1 \left(\nabla_a \bar{\tau}^\delta(a^0, r)^\top v \right) r + \tfrac{1}{2}(s^1)^2 \left(v^\top \nabla_a^2 \bar{\tau}^\delta(a^0, r) v \right) r \right\|.$$

For the solution of the equation

$$\left\| s^1 v + s^1 \nabla_a \bar{\tau}^\delta(a^0, r)^\top v r + \frac{1}{2}(s^1)^2 (v^\top \nabla_a^2 \bar{\tau}^\delta(a^0, r) v) r \right\| = \alpha \quad (4.10)$$

w.r.t. s^1 in general numerical methods as the Newton's method have to be applied. A start point for such a numerical method can be gained by using (4.9). Solving (4.10) is for the Euclidean norm equivalent to find the roots w.r.t. s of the following polynomial:

$$\kappa_4 (s)^4 + \kappa_3 (s)^3 + \kappa_2 (s)^2 + \kappa_0$$

with

$$\kappa_4 = \tfrac{1}{4} \cdot \left(v^\top \nabla_a^2 \bar{\tau}^\delta(a^0, r) v\right)^2 \cdot \|r\|_2^2,$$

$$\kappa_3 = \left(v^\top r + \nabla_a \bar{\tau}^\delta(a^0, r)^\top v \|r\|_2^2\right) \cdot \left(v^\top \nabla_a^2 \bar{\tau}^\delta(a^0, r) v\right),$$

$$\kappa_2 = \|v + \nabla_a \bar{\tau}^\delta(a^0, r)^\top v\, r\|_2^2,$$

$$\kappa_0 = -\alpha^2.$$

In that case, i. e. $K = \mathbb{R}_+^2$, $C = \mathbb{R}_+^p$ and non-degeneracy, we also know (see p. 86) that active constraints remain active. Therefore, if $a^0 + t^0 r - f(x^0) = 0_2$, it also holds $a + t r - f(x) = 0_2$ for a in a neighborhood of a^0 and (t, x) the local minimal solution to the parameter (a, r). Then it is also $(t(a,r), x(a,r)) = (\bar{t}(a,r), \bar{x}(a,r))$ for a in a neighborhood of a^0.

In addition to approximating the minimal value $f(x(a^1, r))$ for determining the new parameter a^1 we can also approximate the minimal solution $(t(a^1, r), x(a^1, r))$ (for the case of the natural ordering, i. e. for $K = \mathbb{R}_+^m$, $C = \mathbb{R}_+^m$). This approximated point can be used as a starting point for a numerical method for solving the problem $(SP(a^1, r))$. Moreover such an approximation of the minimal solution is important for the treatment of multiobjective bilevel optimization problems which is discussed in Chap. 7. There, this is used for determining an equidistant approximation of the set of K-minimal points in the parameter space. In the non-degenerated case such an approximation can be got with the help of Theorem 3.16 and in the degenerated case based on Theorem 3.17. We demonstrate this on a simple example by Hazen ([103, p.186]).

Example 4.7. We consider the unconstrained bicriteria optimization problem

$$\min \begin{pmatrix} \frac{x_1^2}{2} + x_2^2 - 10x_1 - 100 \\ x_1^2 + \frac{x_2^2}{2} - 10x_2 - 100 \end{pmatrix}$$

subject to the constraint

$$x \in \mathbb{R}^2$$

(compare also Test Problem 1 on p.141). For determining efficient points we use the scalar optimization problem $(SP(a, r))$ with

4.2 Parameter Control in the Bicriteria Case

$$a \in H := \{y = (y_1, y_2) \in \mathbb{R}^2 \mid y_2 = 0\}$$

and with constant $r = r^0 = (0,1)^\top$. The parameter $a^0 := (-133.5, 0)^\top$ results in the minimal solution $(t^0, x_1^0, x_2^0) = (-93, 5, 2)$ with Lagrange multiplier $(\mu_1^0, \mu_2^0) = (2, 1)$. It is easy to verify that the assumptions of Theorem 3.13 are satisfied. The derivative of the local minimal value function is thus given by

$$\nabla_a T^\delta(a^0, r^0) = \begin{pmatrix} -2 \\ -1 \end{pmatrix}.$$

If we set now $a^1 := a^0 + s^1 v$ with s^1 as in (4.9) for $v = (1,0)^\top$ and $\alpha = 20$ we obtain $s^1 = 4\sqrt{5}$ and $a^1 = (-133.5 + 4\sqrt{5}, 0)^\top$. For obtaining this result the point $f(x(a^1, r^0))$ is approximated by

$$f(x(a^1, r^0)) \approx f(x^0) + s^1 v + s^1 \left(\nabla_a T^\delta(a^0, r^0)^\top v\right) r^0$$

$$= \begin{pmatrix} -133.5 \\ -93 \end{pmatrix} + 4\sqrt{5} \begin{pmatrix} 1 \\ 0 \end{pmatrix} + 4\sqrt{5} \left(\begin{pmatrix} -2 \\ -1 \end{pmatrix}^\top \begin{pmatrix} 1 \\ 0 \end{pmatrix}\right) \begin{pmatrix} 0 \\ 1 \end{pmatrix}$$

$$= \begin{pmatrix} -124.5557\ldots \\ -110.8885\ldots \end{pmatrix} \tag{4.11}$$

(compare (4.7)).

Using Theorem 3.16 the minimal solution $(t(a^1, r^0), x(a^1, r^0))$ and the Lagrange multipliers $(\mu_1(a^1, r^0), \mu_2(a^1, r^0))$ can also be approximated. It is

$$M = \begin{pmatrix} 0 & 0 & 0 & 0 & -1 \\ 0 & 4 & 0 & -5 & 10 \\ 0 & 0 & 5 & 4 & -8 \\ 0 & 10 & -8 & 0 & 0 \\ 1 & -10 & 8 & 0 & 0 \end{pmatrix} \quad \text{and} \quad N = \begin{pmatrix} 0 & 0 & 0 & -2 & 0 \\ 0 & 0 & 0 & 0 & -1 \\ 0 & 0 & 0 & 186 & 0 \\ 0 & 0 & 0 & 0 & 93 \end{pmatrix}^\top$$

and we obtain the approximation

$$\begin{pmatrix} t(a^1, r^0) \\ x_1(a^1, r^0) \\ x_2(a^1, r^0) \\ \mu_1(a^1, r^0) \\ \mu_2(a^1, r^0) \end{pmatrix} = \begin{pmatrix} -93 \\ 5 \\ 2 \\ 2 \\ 1 \end{pmatrix} + M^{-1} N \begin{pmatrix} 4\sqrt{5} \\ 0 \\ 0 \\ 0 \end{pmatrix} + o\left(\left\| \begin{pmatrix} 4\sqrt{5} \\ 0 \\ 0 \\ 0 \end{pmatrix} \right\|\right)$$

114   4 Adaptive Parameter Control

$$= \begin{pmatrix} -110.8885\ldots \\ 3.8168\ldots \\ 2.7571\ldots \\ 1.0535\ldots \\ 1 \end{pmatrix} + o\left(\left\|\begin{pmatrix} 4\sqrt{5} \\ 0 \\ 0 \\ 0 \end{pmatrix}\right\|\right).$$

With the help of Theorem 3.17 instead of Theorem 3.16 we first have to solve the following system of equations:

$$\begin{pmatrix} 0 & 0 & 0 \\ 0 & 4 & 0 \\ 0 & 0 & 5 \end{pmatrix} \begin{pmatrix} \bar{t} \\ \bar{x} \end{pmatrix} - \bar{\mu}_1 \begin{pmatrix} 0 \\ 5 \\ -4 \end{pmatrix} - \bar{\mu}_2 \begin{pmatrix} 1 \\ -10 \\ 8 \end{pmatrix} = \begin{pmatrix} 0 \\ 0 \\ 0 \end{pmatrix},$$

$$\begin{pmatrix} 0 \\ 5 \\ -4 \end{pmatrix}^\top \begin{pmatrix} \bar{t} \\ \bar{x} \end{pmatrix} = -1,$$

$$\begin{pmatrix} 1 \\ -10 \\ 8 \end{pmatrix}^\top \begin{pmatrix} \bar{t} \\ \bar{x} \end{pmatrix} = 0.$$

The rounded solution is $(\bar{t}, \bar{x}, \bar{\mu}) = (-2, -0.1323, 0.0847, -0.1058, 0)$ and thus we obtain with $a^1 = a^0 + s\,v$, $s = 4\sqrt{5}$:

$$\begin{pmatrix} t(a^1, r^0) \\ x(a^1, r^0) \\ \mu(a^1, r^0) \end{pmatrix} \approx \begin{pmatrix} t^0 \\ x^0 \\ \mu^0 \end{pmatrix} + s \cdot \begin{pmatrix} \bar{t} \\ \bar{x} \\ \bar{\mu} \end{pmatrix} \cdot v = \begin{pmatrix} -110.8885\ldots \\ 3.8166\ldots \\ 2.7575\ldots \\ 1.0536\ldots \\ 1 \end{pmatrix}.$$

For a comparison the minimal solution of the problem $(\mathrm{SP}(a^1, r^0))$ for $a^1 = (-133.5 + 4\sqrt{5}, 0)^\top$ calculated by applying a SQP algorithm (as implemented in Matlab) is

$$\begin{pmatrix} t^1 \\ x_1^1 \\ x_2^1 \end{pmatrix} = \begin{pmatrix} -107.5620\ldots \\ 3.9994\ldots \\ 2.7277\ldots \end{pmatrix}$$

with Lagrange multiplier $(\mu_1^1, \mu_2^1) = (1.3330\ldots, 1)$ and

4.2 Parameter Control in the Bicriteria Case

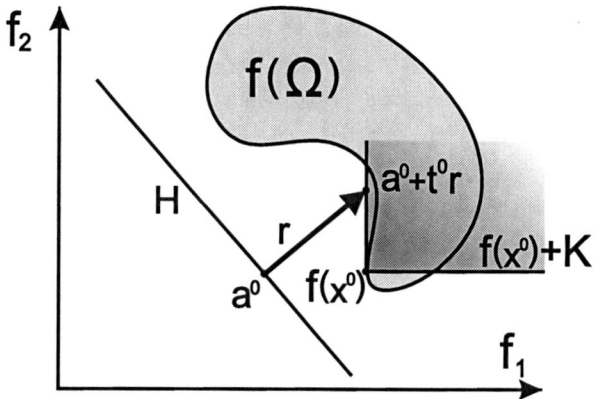

Fig. 4.3. Bicriteria example with $a^0 + t^0 r - f(x^0) \neq 0_2$ for $K = \mathbb{R}^2_+$.

$$f(x^1) = \begin{pmatrix} -124.5557\ldots \\ -107.5620\ldots \end{pmatrix}.$$

The actual distance between the points $f(x^0)$ and $f(x^1)$ is thus

$$\|f(x^1) - f(x^0)\|_2 = 17.0895\ldots.$$

A better result would be achieved by using a Taylor approximation of second or higher order in (4.11) and also by reducing the distance α.

So far we have only discussed the case that the inequality constraint $a^0 + tr - f(x) \in K$ of (SP(a^0, r)) is active in the point (t^0, x^0), i.e. $a^0 + t^0 r - f(x^0) = 0_2$. Now we turn our attention to the case that there is a $k^0 \in K \setminus \{0_2\}$ with

$$a^0 + t^0 r - f(x^0) = k^0$$

as illustrated in Fig. 4.3. Then, according to Theorem 2.21, there is a $\tilde{t}^0 \in \mathbb{R}$ and a parameter $\tilde{a}^0 \in H$ so that

$$\tilde{a}^0 + \tilde{t}^0 r - f(x^0) = 0_2,$$

i.e. the constraint $\tilde{a}^0 + tr - f(x) \in K$ of the problem (SP($\tilde{a}^0, r$)) is active in the point $(\tilde{t}^0, x^0)$. According to Lemma 2.26 the Lagrange multipliers, which are important for determining the derivative of the

116 4 Adaptive Parameter Control

local minimal value function in the point $(\tilde{a}^0, r)$, stay unchanged: if (μ^0, ν^0, ξ^0) are Lagrange multipliers to the minimal solution (t^0, x^0) for $(SP(a^0, r))$, then (μ^0, ν^0, ξ^0) are also Lagrange multipliers to the minimal solution $(\tilde{t}^0, x^0)$ for $(SP(\tilde{a}^0, r))$.

Then we can continue the determination of the next parameter a^1 starting from the point $\tilde{a}^0$ and not from the point a^0, now with the discussed constraint being active. The points $\tilde{t}^0$ and $\tilde{a}^0$ with $\tilde{a}^0 \in H$ can be calculated by

$$\tilde{t}^0 := \frac{b^\top f(x^0) - \beta}{b^\top r} \quad \text{and} \quad \tilde{a}^0 := f(x^0) - \tilde{t}^0 r, \qquad (4.12)$$

see the proof of Theorem 2.21.

The following theorem shows that the line segment with endpoints a^0 and $\tilde{a}^0$ on the hyperplane H can be neglected, and thus that all parameters a with $a = a^0 + \mu(\tilde{a}^0 - a^0) \in H$ for $\mu \in]0, 1[$ can be ignored, without loosing the possibility that all K-minimal points of the multiobjective optimization problem (MOP) can be found by solving $(SP(a, r))$ for appropriate parameters $a \in H$. This fact is even true for the case of more than two objective functions, i.e. for $m \geq 3$. We use for this result that according to Theorem 2.17 for any K-minimal point $\bar{x} \in \mathcal{M}(f(\Omega), K)$ there exists a parameter $\bar{a} \in H^a$ and some $\bar{t} \in \mathbb{R}$ so that $(\bar{t}, \bar{x})$ is a minimal solution of $(SP(\bar{a}, r))$ with $\bar{a} + \bar{t} r - f(\bar{x}) = 0_m$. The points $\bar{t}$ and $\bar{a}$ are given by

$$\bar{t} = \frac{b^\top f(\bar{x}) - \beta}{b^\top r} \quad \text{and} \quad \bar{a} = f(\bar{x}) - \bar{t} r. \qquad (4.13)$$

Theorem 4.8. *Let (t^0, x^0) be a minimal solution of $(SP(a^0, r))$ for a parameter $a^0 \in H = \{y \in \mathbb{R}^m \mid b^\top y = \beta\}$ with*

$$a^0 + t^0 r - f(x^0) = k^0 \qquad (4.14)$$

and $k^0 \in K \setminus \{0_m\}$. Then there exists no K-minimal point $\bar{x} \in \mathcal{M}(f(\Omega), K)$ with

$$\bar{a} = a^0 + \mu(\tilde{a}^0 - a^0) \quad \text{for some } \mu \in]0, 1[\qquad (4.15)$$

and $\bar{a}$ as in (4.13) and $\tilde{a}^0$ as in (4.12).

Proof. We proof this by contradiction and assume that there exists a K-minimal point $\bar{x}$ with (4.15). We set

4.2 Parameter Control in the Bicriteria Case

$$t' := \bar{t} - (1-\mu)(t^0 - \tilde{t}^0)$$

and do a case differentiation.

With (4.12) and (4.14) we have

$$\begin{aligned}\tilde{a}^0 - a^0 &= f(x^0) - \tilde{t}^0 r - (k^0 - t^0 r + f(x^0)) \\ &= (t^0 - \tilde{t}^0)r - k^0.\end{aligned} \quad (4.16)$$

Case $t' \geq \tilde{t}^0$. With (4.13), (4.14), (4.15), (4.16) and the definition of t' we conclude for an arbitrary $\mu \in]0,1[$

$$\begin{aligned}f(\bar{x}) - f(x^0) &= \bar{a} + \bar{t}r - (a^0 + t^0 r - k^0) \\ &= a^0 + \mu(\tilde{a}^0 - a^0) + \bar{t}r - (a^0 + t^0 r - k^0) \\ &= \mu((t^0 - \tilde{t}^0)r - k^0) + (\bar{t} - t^0)r + k^0 \\ &= \mu((t^0 - \tilde{t}^0)r - k^0) + (t' + (1-\mu)(t^0 - \tilde{t}^0) - t^0)r + k^0 \\ &= \underbrace{(1-\mu)}_{>0}\underbrace{k^0}_{\neq 0_m} + \underbrace{(t' - \tilde{t}^0)}_{\geq 0} r \in K \setminus \{0_m\}\end{aligned}$$

in contradiction to the K-minimality of $\bar{x}$.

Case $t' < \tilde{t}^0$. We first show that the point $(t^0 - \tilde{t}^0 + t', \bar{x})$ is feasible for the problem $(\mathrm{SP}(a^0, r))$. Using the definition of t' together with (4.13), (4.15) and (4.16) we obtain for $\mu \in]0,1[$

$$\begin{aligned}a^0 + (t^0 - \tilde{t}^0 + t')r - f(\bar{x}) &= a^0 + \left(t^0 - \tilde{t}^0 + \bar{t} - (1-\mu)(t^0 - \tilde{t}^0)\right)r \\ &\quad -(\bar{a} + \bar{t}r) \\ &= (a^0 - \bar{a}) + \mu(t^0 - \tilde{t}^0)r \\ &= -\mu(\tilde{a}^0 - a^0) + \mu(t^0 - \tilde{t}^0)r \\ &= -\mu((t^0 - \tilde{t}^0)r - k^0) + \mu(t^0 - \tilde{t}^0)r \\ &= \mu k^0 \in K.\end{aligned}$$

Due to $t' < \tilde{t}^0$ it is $t^0 - \tilde{t}^0 + t' = t^0 + (t' - \tilde{t}^0) < t^0$ and thus this is a contradiction to (t^0, x^0) a minimal solution of $(\mathrm{SP}(a^0, r))$. □

As preconcerted we determine the parameters a in increasing order w.r.t. the first coordinate, i.e. $a_1^0 \leq a_1^1 \leq a_1^2 \leq \ldots$. Thus we are interested in the question whether it is $\tilde{a}_1^0 > a_1^0$ or $\tilde{a}_1^0 \leq a_1^0$. For the following examinations we restrict ourselves to the special case of the natural ordering $K = \mathbb{R}_+^2$ in the image space. We generally assume that a hyperplane $H = \{y \in \mathbb{R}^2 \mid b^\top y = \beta\}$ is given and the parameter $r \in \mathbb{R}_+^2$ with $b^\top r \neq 0$ is assumed to be constant.

4 Adaptive Parameter Control

Lemma 4.9. *Let* $K = \mathbb{R}_+^2$. *Let* (t^0, x^0) *be a minimal solution of* $(SP(a^0, r))$ *for a parameter* $a^0 \in H = \{y \in \mathbb{R}^2 \mid b^\top y = \beta\}$ *with*

$$a^0 + t^0 r - f(x^0) = k^0 \in \mathbb{R}_+^2 \setminus \{0_2\}. \qquad (4.17)$$

Then for $\tilde{a}^0$ *as in (4.12) it is* $\tilde{a}_1^0 > a_1^0$ *if and only if*

$$(k_1^0 = 0, \ k_2^0 > 0 \ \text{and} \ \tfrac{r_1 b_2}{b^\top r} > 0) \ \text{or}$$

$$(k_1^0 > 0, \ k_2^0 = 0 \ \text{and} \ \tfrac{r_2 b_2}{b^\top r} < 0).$$

Proof. According to Theorem 2.1,c) it is $k^0 \in \partial \mathbb{R}_+^2$. We have with (4.12) and (4.17)

$$\tilde{a}_1^0 > a_1^0$$

$\Leftrightarrow \qquad f_1(x^0) - \tilde{t}^0 r_1 > k_1^0 - t^0 r_1 + f_1(x^0)$

$\Leftrightarrow \qquad (t^0 - \tilde{t}^0) r_1 > k_1^0$

$\Leftrightarrow \qquad \frac{(b^\top (f(x^0)+k^0)-\beta)-(b^\top f(x^0)-\beta)}{b^\top r} r_1 > k_1^0$

$\Leftrightarrow \qquad \frac{b^\top k^0}{b^\top r} r_1 > k_1^0$

$\Leftrightarrow \qquad \frac{r_1 b_2}{b^\top r} k_2^0 > \left(1 - \frac{r_1 b_1}{b^\top r}\right) k_1^0.$

This inequality is for $k^0 \in \partial \mathbb{R}_+^2$ satisfied if and only if

$$k_1^0 = 0, \ k_2^0 > 0 \ \text{and} \ \tfrac{r_1 b_2}{b^\top r} > 0 \ \text{or}$$

$$k_1^0 > 0, \ k_2^0 = 0 \ \text{and} \ 1 - \tfrac{r_1 b_1}{b^\top r} = \tfrac{r_2 b_2}{b^\top r} < 0.$$

□

This result is illustrated by the following example.

Example 4.10. We consider the bicriteria optimization problem

$$\min \ f(x) = x$$

subject to the constraint

$$(x_1 - 4)^2 + (x_2 - 1)^2 \leq 1,$$

$$x \in \mathbb{R}^2$$

w. r. t. the natural ordering. The efficient set is given by

4.2 Parameter Control in the Bicriteria Case

$$\mathcal{E}(f(\Omega), \mathbb{R}_+^2) = \{y \in \mathbb{R}^2 \mid (y_1 - 4)^2 + (y_2 - 1)^2 = 1,\ y_1 \leq 4,\ y_2 \leq 1\}.$$

Let the hyperplane H be defined by $H := \{y \in \mathbb{R}^2 \mid (1,1)y = 0\}$, i.e. $b = (1,1)^\top$, $\beta = 0$, and let $r := (1,1)^\top$. For $a^0 := (0,0)$ the minimal solution of the Pascoletti-Serafini scalarization to the considered bicriteria optimization problem is $(t^0, x^0) = (3, 3, 1)^\top$, with

$$k^0 := a^0 + t^0 r - f(x^0) = (0, 2)^\top.$$

We calculate $\tilde{a}^0$ by

$$\tilde{a}^0 := f(x^0) - \tilde{t}^0 r = (1, -1)^\top \text{ with } \tilde{t}^0 := \frac{1}{b^\top r}(b^\top f(x^0) - \beta) = 2$$

(see Fig. 4.4). It is $\tilde{a}_1^0 = 1 > a_1^0 = 0$ and $k_1^0 = 0$, $k_2^0 > 0$, $\frac{r_2 b_2}{b^\top r} = \frac{1}{2} > 0$, confirming the result of Lemma 4.9.

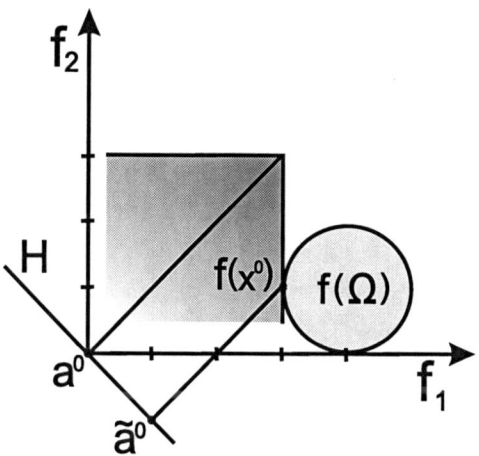

Fig. 4.4. Illustration of Example 4.10.

Summarizing this we proceed in the case that $k^0 = a^0 + t^0 r - f(x^0) \neq 0_2$ for (t^0, x^0) a minimal solution of $(SP(a^0, r))$ as follows: first we determine the new parameter $\tilde{a}^0$ by

$$\tilde{a}^0 := f(x^0) - \tilde{t}^0 r \text{ with } \tilde{t}^0 := \frac{b^\top f(x^0) - \beta}{b^\top r}.$$

According to Theorem 4.8 the parameters $a \in \{a^0 + \mu \cdot (\tilde{a}^0 - a^0) \mid \mu \in\ ]0, 1[\}$ can be neglected.

120   4 Adaptive Parameter Control

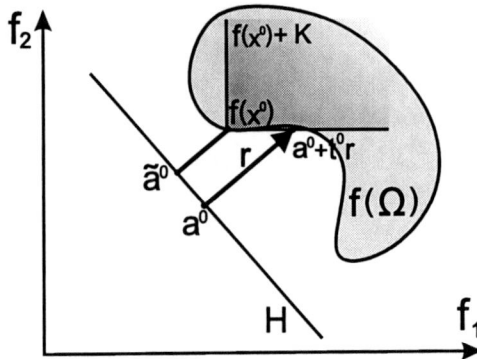

Fig. 4.5. Bicriteria example with $a^0 + t^0 r - f(x^0) \neq 0_2$ for $K = \mathbb{R}_+^2$ and $\tilde{a}_1^0 < a_1^0$.

A criteria for $\tilde{a}_1^0 > a_1^0$ is given in Lemma 4.9. If this is the case we continue the parameter determination with the parameter $\tilde{a}^0$. As already mentioned (Lemma 2.26) the point $(\tilde{t}^0, x^0)$ is a minimal solution of (SP$(\tilde{a}^0, r)$) with $\tilde{a}^0 + \tilde{t}^0 r - f(x^0) = 0_2$ and the Lagrange multipliers are known.

For the case $\tilde{a}_1^0 \leq a_1^0$ as illustrated in Fig. 4.5 it is not useful to continue with the parameter $\tilde{a}^0$ instead of a^0 as we are looking for parameters with increasing first coordinate. In that case we still use the parameter a^0 for determining a^1. We can no longer assume $f(x(a,r)) = a + t(a,r)r$ as we have $f(x^0) = a^0 + t^0 r - k^0$ with $k^0 \neq 0_2$. However we can assume that the constraint $a + tr - f(x) \in K$ remains inactive and thus in view of $a^0 + t^0 r = f(x^0) + k^0$ we assume

$$a + t(a,r) r = f(x^0) + k^0 + s k^0.$$

Then, for

$$s = \frac{\alpha}{\|k^0\|},$$

we have a distance of $\alpha > 0$ between the points $a+t(a,r)r$ and $a^0+t^0 r$, see Fig. 4.6. Thus we set for the new parameter

$$a^1 := f(x^0) + (1+s) k^0 - t r$$

with $s = \alpha/\|k^0\|$ and for some $t \in \mathbb{R}$. As we still demand $a \in H$ we choose

$$t := \frac{b^\top \left(f(x^0) + (1+s) k^0 \right) - \beta}{b^\top r}. \tag{4.18}$$

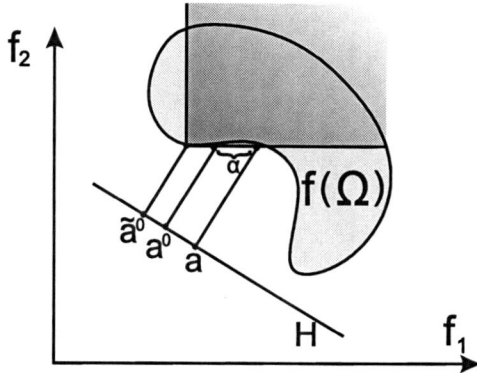

Fig. 4.6. Choosing the new parameter if $a^0 + t^0 r - f(x^0) \neq 0_2$ and $\tilde{a}_1^0 < a_1^0$.

Using the definition of $\tilde{a}^0$ (see (4.12)) this results in

$$a^1 = f(x^0) + (1+s)k^0 - \frac{b^\top \left(f(x^0) + (1+s)k^0\right) - \beta}{b^\top r} r$$

$$= \tilde{a}^0 + (1+s)\left(k^0 - \frac{b^\top k^0}{b^\top r} r\right).$$

According to

$$b^\top a^1 = \underbrace{b^\top \tilde{a}^0}_{=\beta} + (1+s)\underbrace{\left(b^\top k^0 - \frac{b^\top k^0}{b^\top r} b^\top r\right)}_{=0} = \beta,$$

the parameter a^1 is an element of the hyperplane H. Further we predefined that the parameters a are increasing w. r. t. their first coordinate. This is the case as shown in the following lemma:

Lemma 4.11. *Let $K = \mathbb{R}_+^2$. Let (t^0, x^0) be a minimal solution of $(SP(a^0, r))$ for a parameter $a^0 \in H = \{y \in \mathbb{R}^2 \mid b^\top y = \beta\}$ with*

$$a^0 + t^0 r - f(x^0) = k^0 \in \mathbb{R}_+^2 \setminus \{0_2\}. \tag{4.19}$$

Then for $\tilde{a}^0$ as in (4.12) let $\tilde{a}_1^0 \leq a_1^0$ and we set

$$a := \tilde{a}^0 + (1+s)\left(k^0 - \frac{b^\top k^0}{b^\top r} r\right) \tag{4.20}$$

for some $s > 0$. Then $a_1 \geq a_1^0$.

4 Adaptive Parameter Control

Proof. Using the definition of $\tilde{a}^0$ we have

$$a = \tilde{a}^0 + (1+s)\left(k^0 - \frac{b^\top k^0}{b^\top r}r\right)$$
$$= f(x^0) + (1+s)\,k^0 - t\,r \qquad (4.21)$$

with t as in (4.18). If we notice that

$$t^0 = \frac{b^\top(f(x^0) + k^0) - \beta}{b^\top r}$$

we get

$$t - t^0 = \frac{1}{b^\top r}\left(b^\top f(x^0) + (1+s)b^\top k^0 - \beta - b^\top f(x^0) - b^\top k^0 + \beta\right)$$
$$= \frac{s}{b^\top r}b^\top k^0. \qquad (4.22)$$

Using (4.21), (4.19) and (4.22) the inequality $a_1 \geq a_1^0$ is equivalent to

$$f_1(x^0) + (1+s)k_1^0 - t\,r_1 \geq f_1(x^0) + k_1^0 - t^0\,r_1$$
$$\Leftrightarrow \quad sk_1^0 - (t - t^0)r_1 \geq 0$$
$$\Leftrightarrow \quad s\left(k_1^0 - \frac{b^\top k^0}{b^\top r}r_1\right) \geq 0. \qquad (4.23)$$

According to the assumptions of this lemma it is $k^0 \neq 0_2$ and with Theorem 2.1,c) we conclude $k^0 \in \partial\mathbb{R}^2_+ \setminus \{0_2\}$. Thus we have $k_1^0 > 0$, $k_2^0 = 0$ or $k_1^0 = 0$, $k_2^0 > 0$.

Case $k_1^0 > 0$, $k_2^0 = 0$. Then the inequality (4.23) is equivalent to

$$s\left(k_1^0 - \frac{b_1 k_1^0}{b^\top r}r_1\right) = \underbrace{sk_1^0}_{>0}\left(1 - \frac{b_1 r_1}{b^\top r}\right) \geq 0.$$

We have $\tilde{a}_1^0 \leq a_1^0$ and thus with Lemma 4.9 we conclude

$$\frac{r_2 b_2}{b^\top r} = 1 - \frac{b_1 r_1}{b^\top r} \geq 0.$$

Hence the inequality (4.23) is satisfied, i.e. $a_1 \geq a_1^0$.

Case $k_1^0 = 0$, $k_2^0 > 0$. Again applying Lemma 4.9 we obtain

$$\frac{r_1 b_2}{b^\top r} \leq 0$$

4.2 Parameter Control in the Bicriteria Case

and thus (4.23) is equivalent to

$$-s\frac{b_2 k_2^0}{b^\top r} r_1 \geq 0$$

$$\Leftrightarrow \quad \underbrace{-sk_2^0}_{<0} \underbrace{\frac{b_2 r_1}{b^\top r}}_{\leq 0} \geq 0$$

and is satisfied. This completes the proof. □

So, in the case $a^0 + t^0 r - f(x^0) = k^0 \neq 0_2$ with $\tilde{a}_1^0 \leq a_1^0$ we choose the next parameter a^1 by

$$a^1 := \tilde{a}^0 + \left(1 + \frac{\alpha}{\|k^0\|}\right)\left(k^0 - \frac{b^\top k^0}{b^\top r} r\right)$$

and according to Lemma 4.11 it is $a_1^1 \geq a_1^0$.

4.2.1 Algorithm for the Pascoletti-Serafini Scalarization

We now summarize the discussed results and formulate an algorithm for solving bicriteria optimization problems w.r.t. the natural ordering based on the Pascoletti-Serafini scalarization. For that let the Assumption 3.12 as well as the Assumption 4.6 for $m = 2$ be satisfied. In detail we assume the following:

Assumption 4.12. Let $m = 2$, $K = \mathbb{R}_+^2$, $C = \mathbb{R}_+^p$ and $\hat{S} = S = \mathbb{R}^n$. Let the functions $f\colon \mathbb{R}^n \to \mathbb{R}^2$, $g\colon \mathbb{R}^n \to \mathbb{R}^p$, and $h\colon \mathbb{R}^n \to \mathbb{R}^q$ be twice continuously differentiable on $\mathbb{R}^n$. Further, to any parameters a and r for which we consider the optimization problem (SP(a,r)) (and ($\overline{\text{SP}}(a,r)$) respectively) let there exist a minimal solution $(\bar{t}, \bar{x})$ with Lagrange multipliers $(\bar{\mu}, \bar{\nu}, \bar{\xi}) \in \mathbb{R}_+^m \times \mathbb{R}_+^p \times \mathbb{R}^q$ and let the assumptions of Theorem 3.13 (and Theorem 3.7 respectively) be satisfied in $(\bar{t}, \bar{x})$.

Thus we consider only multiobjective optimization problems which can be formulated as

$$\min f(x) = \begin{pmatrix} f_1(x) \\ f_2(x) \end{pmatrix}$$

subject to the constraints

$$g_j(x) \geq 0, \quad j = 1, \ldots, p,$$
$$h_k(x) = 0, \quad k = 1, \ldots, q,$$
$$x \in \mathbb{R}^n$$

4 Adaptive Parameter Control

with the image space $\mathbb{R}^2$ partially ordered by the natural ordering. The constraint set is

$$\Omega = \{x \in \mathbb{R}^n \mid g_j(x) \geq 0, \ j = 1, \ldots, p, \ h_k(x) = 0, \ k = 1, \ldots, q\}.$$

We choose the parameter $r \in K = \mathbb{R}^2_+$ constant in the Pascoletti-Serafini scalarization and determine an approximation of the efficient set by varying the parameter a. Without loss of generality we assume $r_1 > 0$. According to Theorem 3.13 and Lemma 3.14 we get for the local minimal value function in any point $(a^0, r) \in \mathbb{R}^2 \times \mathbb{R}^2$

$$\nabla_a \tau^\delta(a^0, r) = -\mu^0$$

with $\mu^0 \in \mathbb{R}^2_+$ the Lagrange multiplier to the constraint $a^0 + tr - f(x) \geq_2 0_2$.

Using this information we formulate the following algorithm.

Algorithm 1 (Adaptive Pascoletti-Serafini method for bicriteria problems).

Input: Choose $r \in \mathbb{R}^2_+$, with $r_1 > 0$, and predefine the desired distance $\alpha \in \mathbb{R}$, $\alpha > 0$, between the approximation points. Choose a hyperplane

$$H = \{y \in \mathbb{R}^2 \mid b^\top y = \beta\}$$

by setting $b \in \mathbb{R}^2$ with $b^\top r \neq 0$ and $\beta \in \{0, 1\}$. Deliver $M^1 \in \mathbb{R}$ with

$$M^1 > f_2(x) - f_1(x) \frac{r_2}{r_1} \quad \text{for all} \ x \in \Omega.$$

Step 1: Solve (SP($\tilde{a}^1, r$)) for $\tilde{a}^1 = (0, M^1)^\top$ with minimal solution $(\tilde{t}^1, x^1)$ and Lagrange multiplier $\mu^1 \in \mathbb{R}^2_+$ to the constraint $\tilde{a}^1 + tr - f(x) \geq_2 0_2$. Calculate

$$t^1 := \frac{b^\top f(x^1) - \beta}{b^\top r} \quad \text{and} \quad a^1 := f(x^1) - t^1 r.$$

Set $k^1 := 0_2$, $l := 1$.

4.2 Parameter Control in the Bicriteria Case

Step 2: Solve $\min_{x \in \Omega} f_2(x)$ with minimal solution x^E and calculate

$$t^E := \frac{b^\top f(x^E) - \beta}{b^\top r} \quad \text{and} \quad a^E := f(x^E) - t^E r.$$

Set $v := a^E - a^1$.

Step 3: Determine a^{l+1} by:
- If $k^l = 0_2$ set

$$a^{l+1} := a^l + \frac{\alpha}{\|v + ((-\mu^l)^\top v)\, r\|} \cdot v. \qquad (4.24)$$

- Elseif $(k_1^l = 0,\ k_2^l > 0 \quad \text{and} \quad \frac{r_1 b_2}{b^\top r} > 0)$ or $(k_1^l > 0,\ k_2^l = 0 \quad \text{and} \quad \frac{b_2 r_2}{b^\top r} < 0)$ calculate

$$\tilde{t}^l := \frac{b^\top f(x^l) - \beta}{b^\top r} \quad \text{and} \quad \tilde{a}^l := f(x^l) - \tilde{t}^l r$$

and set

$$a^{l+1} := \tilde{a}^l + \frac{\alpha}{\|v + ((-\mu^l)^\top v)\, r\|} \cdot v.$$

- Elseif $(k_1^l = 0,\ k_2^l > 0 \quad \text{and} \quad \frac{r_1 b_2}{b^\top r} \leq 0)$ or $(k_1^l > 0,\ k_2^l = 0 \quad \text{and} \quad \frac{b_2 r_2}{b^\top r} \geq 0)$ set

$$a^{l+1} := \tilde{a}^l + \left(1 + \frac{\alpha}{\|k^l\|}\right)\left(k^l - \frac{b^\top k^l}{b^\top r} r\right).$$

Step 4: Set l:=l+1.
If $a^l = a^1 + \lambda \cdot v$ for a $\lambda \in [0,1]$, solve (SP(a^l, r)) with minimal solution (t^l, x^l) and Lagrange multiplier μ^l to the constraint $a^l + tr - f(x) \geq_2 0_2$, set $k^l := a^l + t^l r - f(x^l)$ and go to Step 3.
Else stop.

Output: The set $A := \{x^1, \cdots, x^{l-1}, x^E\}$ is an approximation of the set $\mathcal{M}_w(f(\Omega), K)$.

In the Input it is asked for a value $M^1 \in \mathbb{R}$ with $M^1 > f_2(x) - f_1(x)\frac{r_2}{r_1}$. Such a point exists if there exist minimal solutions of the problems $\max_{x \in \Omega} f_2(x)$ and $\min_{x \in \Omega} f_1(x)$. Then we can choose

$$M^1 > \max_{x \in \Omega} f_2(x) - \min_{x \in \Omega} f_1(x)\frac{r_2}{r_1}.$$

The optimization problem $(\mathrm{SP}(\tilde{a}^1, r))$ in Step 1 is

$$\min t$$

subject to the constraints

$$t r_1 - f_1(x) \geq 0, \qquad (4.25)$$
$$M^1 + t r_2 - f_2(x) \geq 0,$$
$$t \in \mathbb{R}, \ x \in \Omega.$$

Thus, for any feasible point (t, x) it holds $t \geq \frac{f_1(x)}{r_1}$ and then the second inequality

$$M^1 + t r_2 - f_2(x) > \left(f_2(x) - f_1(x)\frac{r_2}{r_1}\right) + \frac{f_1(x)}{r_1}r_2 - f_2(x) = 0$$

is always satisfied. So, we can replace problem (4.25) by

$$\min \frac{f_1(x)}{r_1}$$

subject to the constraint

$$x \in \Omega$$

being equivalent to $\min_{x \in \Omega} f_1(x)$. We have $\min_{x \in \Omega} f_1(x) = f_1(x^1)$ and (t^1, x^1) is a minimal solution of $(\mathrm{SP}(a^1, r))$ with Lagrange multiplier μ^1 to the constraint $a^1 + tr - f(x) \geq_2 0_2$. In Step 1 and Step 2 we determine the subset H^a of the hyperplane and we define a direction v with $a^1 + sv \in H$ for all $s \in \mathbb{R}$. Note that generally it is $\tilde{a}^1 \notin H$.

In Step 3 the adaptive parameter control which leads to an approximation A of the set of weakly EP-minimal points is done. We do a case differentiation if the constraint $a^l + tr - f(x) \geq_2 0_2$ is active or not, i.e. if

$$k^l = a^l + t^l r - f(x^l) \neq 0_2$$

or not, and in the case $k^l \neq 0_2$ if it is also $\tilde{a}^l_1 < a^l_1$ or not (see Lemma 4.9).

4.2 Parameter Control in the Bicriteria Case

For solving the scalar optimization problems in the Steps 1, 2 and 4 a numerical method has to be used as for instance the SQP (sequential quadratic programming) method (see among others in [80, 183, 199]). However using just a local solver can lead to only local minimal solutions of the scalar problems and thus to only locally weakly EP-minimal points. Besides it can happen that parts of the efficient set are neglected as it is demonstrated in test problem 4 in Chap. 5. In that case a global solution method has to be applied. For a survey on global optimization methods see [17, 115, 225]. Examples for a global solution method are the method by Schäffler ([192, 193]) based on stochastic differential equations or the method called Direct by Jones, Perttunen and Stuckman ([132]).

As a starting point for the numerical method for solving the problem (SP(a^l, r)) in Step 4 the point (t^{l-1}, x^{l-1}) can be used. For improving this starting point an approximation of the point (t^l, x^l) according to Theorem 3.16 and Theorem 3.17 respectively can be done. The condition $a^l = a^1 + \lambda v$ for a $\lambda \in [0, 1]$ in Step 4 corresponds due to $v = a^E - a^1$ to the condition that a^l is an element of the set H^a, i.e. of the line segment with endpoints a^1 and a^E.

As a result of Algorithm 1 we get an approximation of the set of weakly EP-minimal points. If we choose from the set A the non-dominated points we get an approximation of the set of EP-minimal points. However note that this approximation need not to consist of EP-minimal points only and that it can include just weakly EP-minimal points, too.

The approximation quality can be improved by using second order information. If the constraints are non-degenerated in addition to the other assumptions we have

$$\nabla_a^2 \tau^\delta(a^0, r) = -\nabla_a \mu(a^0, r)$$

according to Theorem 3.15. Here, $\nabla_a^2 \tau^\delta(a^0, r)$ can be approximated by numerical differentiation. For example for $b_1, b_2 \neq 0$ we set

$$\nabla_a^2 \tau^\delta(a^l, r) \approx \begin{pmatrix} \frac{\mu_1^l - \mu_1^{l-1}}{a_1^l - a_1^{l-1}} & \frac{\mu_1^l - \mu_1^{l-1}}{a_2^l - a_2^{l-1}} \\ \frac{\mu_2^l - \mu_2^{l-1}}{a_1^l - a_1^{l-1}} & \frac{\mu_2^l - \mu_2^{l-1}}{a_2^l - a_2^{l-1}} \end{pmatrix} =: H_\tau.$$

Then, a solution $s > 0$ of the equation

$$\| s v - s(\mu^l)^\top v r + \frac{1}{2} s^2 (v^\top H_\tau v) r \|^2 = \alpha^2$$

has to be found (using numerical methods as e. g. the Newton's method). Finally, we choose for the new parameter

$$a^{l+1} := a^l + s \cdot v.$$

4.2.2 Algorithm for the ε-Constraint Scalarization

According to Sect. 2.5.1 the ε-constraint scalarization is a special case of the Pascoletti-Serafini scalarization. Thus we can apply the results for the adaptive parameter control gained for the general method also for controlling the parameter ε. Here, we consider again the bicriteria case only. For three or more objective functions see Sect. 4.3.

We arbitrarily choose $k = 2$ and consider the problem $(P_2(\varepsilon))$

$$\min\ f_2(x)$$
subject to the constraints
$$f_1(x) \leq \varepsilon,$$
$$x \in \Omega.$$

In comparison to the Pascoletti-Serafini problem it is $H = \{y \in \mathbb{R}^2 \mid y_2 = 0\}$, i.e. $b = (0,1)^\top$, $\beta = 0$, and $r = (0,1)^\top$ predefined. We choose $v := (1,0)^\top$. Let the Assumption 4.12 hold. For any parameter $a^0 = (\varepsilon^0, 0)^\top$ the inequality constraint $a_2^0 + t\, r_2 - f_2(x) \geq 0$ is active. Thus, for any minimal solution (t^0, x^0) of the related problem $(SP(a^0, r))$ we have

$$k^0 = a^0 + t^0\, r - f(x^0)$$
$$= \begin{pmatrix} \varepsilon^0 \\ 0 \end{pmatrix} + t^0 \begin{pmatrix} 0 \\ 1 \end{pmatrix} - \begin{pmatrix} f_1(x^0) \\ f_2(x^0) \end{pmatrix}$$
$$= \begin{pmatrix} \varepsilon^0 - f_1(x^0) \\ 0 \end{pmatrix}$$
$$\geq 0_2$$

and hence $k^0 \neq 0_2$ if and only if $k_2^0 = 0$ and $k_1^0 > 0$. Determining $\tilde{a}^0$ as in Lemma 4.9 (see also (4.12)) results in

$$\tilde{a}^0 = \begin{pmatrix} \tilde{\varepsilon}^0 \\ 0 \end{pmatrix} = \begin{pmatrix} f_1(x^0) \\ f_2(x^0) \end{pmatrix} - f_2(x^0) \cdot \begin{pmatrix} 0 \\ 1 \end{pmatrix} = \begin{pmatrix} f_1(x^0) \\ 0 \end{pmatrix}.$$

4.2 Parameter Control in the Bicriteria Case

As a consequence (also as a result of Lemma 4.9) it is never $\tilde{a}_1^0 = f_1(x^0) > a_1^0 = \varepsilon^0$. Further note, that in the case of $\varepsilon^0 - f_1(x^0) = k_1^0 > 0$ the correspondent Lagrange multiplier μ_1^0 is equal to zero. In the following we determine the parameters in decreasing order w.r.t. the first coordinate $a_1 = \varepsilon$ in contrast to the previous procedure.

For $k^0 \neq 0_2$ we choose (compare (4.24) in Step 3 of Alg. 1)

$$a^1 = \begin{pmatrix} \varepsilon^1 \\ 0 \end{pmatrix} = a^0 - \frac{\alpha}{\|v + ((-\mu^0)^\top v)r\|}v$$

$$= \begin{pmatrix} \varepsilon^0 \\ 0 \end{pmatrix} - \frac{\alpha}{\sqrt{1 + (\mu_1^0)^2}}\begin{pmatrix} 1 \\ 0 \end{pmatrix},$$

i.e.

$$\varepsilon^1 = \varepsilon^0 - \frac{\alpha}{\sqrt{1 + (\hat{\mu}^0)^2}}$$

with $\hat{\mu}^0 := \mu_1^0 \in \mathbb{R}^+$ the Lagrange multiplier to the constraint $f_1(x) - \varepsilon^0 \leq 0$. Because $\varepsilon^0 = f_1(x^0)$ we can also write

$$\varepsilon^1 = f_1(x^0) - \frac{\alpha}{\sqrt{1 + (\hat{\mu}^0)^2}}. \tag{4.26}$$

For $k^0 \neq 0_2$, i.e. for $\varepsilon^0 - f_1(x^0) > 0$ and thus $\hat{\mu}^0 = 0$ we continue with the parameter $\tilde{\varepsilon}^0 = f_1(x^0)$ and we choose

$$\varepsilon^1 = f_1(x^0) - \alpha.$$

Note, that due to $\hat{\mu}^0 = 0$ this case is included in the equation (4.26). This results in the following algorithm:

Algorithm 2 (Adaptive ε-constraint method for bicriteria problems).

Input: Choose a desired distance $\alpha > 0$ between the approximation points. Choose $M > f_1(x)$ for all $x \in \Omega$.

Step 1: Solve problem $(P_2(\varepsilon))$ with parameter $\varepsilon := M$ with minimal solution x^1 and Lagrange multiplier μ^1 to the constraint $f_1(x) - \varepsilon^1 \leq 0$. Set $\varepsilon^1 := f_1(x^1)$ and $l := 1$.

Step 2: Solve $\min_{x \in \Omega} f_1(x)$ with minimal solution x^E.

Step 3: Set
$$\varepsilon^{l+1} := f_1(x^l) - \frac{\alpha}{\sqrt{1+(\mu^l)^2}}.$$

Step 4: Set $l := l+1$.
If $\varepsilon^l \geq f_1(x^E)$ solve problem $(P_2(\varepsilon^l))$ with minimal solution x^l and Lagrange multiplier μ^l to the constraint $f_1(x) - \varepsilon^l \leq 0$ and go to Step 3. Otherwise stop.

Output: The set $A := \{x^1, \ldots, x^{l-1}, x^E\}$ is an approximation of the set of weakly EP-minimal points.

An algorithm specialized to the problem $(P_1(\varepsilon))$ can be found in [69], but note, that the problems $(P_1(\varepsilon))$ and $(P_2(\varepsilon))$ are equal just by renaming the functions f_1 and f_2.

We can use again second order sensitivity information. Assuming all necessary assumptions are satisfied we set in Step 3

$$H := \frac{\mu^l - \mu^{l-1}}{\varepsilon^l - \varepsilon^{l-1}}$$

and calculate a solution $\bar{s} < 0$ of the equation

$$s^4(\frac{1}{4}H^2) - s^3(H\mu^l) + s^2(1+(\mu^l)^2) - \alpha^2 = 0$$

e.g. with Newton's method with a starting value of $s = -\frac{\alpha}{\sqrt{1+(\mu^l)^2}}$ and set

$$\varepsilon^{l+1} := \varepsilon^l + \bar{s}$$

(compare [67]).

4.2.3 Algorithm for the Normal Boundary Intersection Scalarization

Next we specialize the general algorithm Alg. 1 to the normal boundary intersection method by Das and Dennis ([38, 40]) which we already

4.2 Parameter Control in the Bicriteria Case

discussed in Sect. 2.5.2. We recall the problem $(NBI(\beta))$:

$$\max s$$
subject to the constraints
$$\Phi\beta + s\bar{n} = f(x) - f^*,$$
$$s \in \mathbb{R}, \ x \in \Omega.$$

In Sect. 2.5.2, Lemma 2.32, we have already seen how the parameter β of this method corresponds to the parameter a of the modified Pascoletti-Serafini scalarization $(\overline{SP}(a,r))$:

$$a = f^* + \Phi\beta.$$

According to Sect. 2.5.2 all parameters $a = f^* + \Phi\beta$ are elements of a hyperplane H with

$$H = \{\beta f(x^1) + (1-\beta) f(x^2) \mid \beta \in \mathbb{R}\}$$

and the set H^a is

$$H^a = \{\beta f(x^1) + (1-\beta) f(x^2) \mid \beta \in [0,1]\}.$$

As direction v (compare Step 2 of Alg. 1) we choose again $v := a^E - a^1$. Here, as we have $a^E = f^* + \Phi\beta^E$, $a^1 = f^* + \Phi\beta^1$ and $\beta^1 = (1,0)^\top$, $\beta^E = (0,1)^\top$ we conclude

$$v = \Phi(\beta^E - \beta^1) = \Phi \begin{pmatrix} -1 \\ 1 \end{pmatrix}.$$

Further, we set $r = -\bar{n}$ with $\bar{n}$ the normal unit vector to the hyperplane H which directs to the negative orthant. Using $a^l = f^* + \Phi\beta^l$ for arbitrary $l \in \mathbb{N}$ we set correspondent to (4.24)

$$f^* + \Phi\beta^{l+1} := f^* + \Phi\beta^l + \frac{\alpha}{\|v + ((\mu^l)^\top v)\bar{n}\|} \Phi \begin{pmatrix} -1 \\ 1 \end{pmatrix}$$

resulting in

$$\beta^{l+1} := \beta^l + \frac{\alpha}{\|v + ((\mu^l)^\top v)\bar{n}\|} \begin{pmatrix} -1 \\ 1 \end{pmatrix}.$$

Based on that we get the following algorithm for controlling the parameter $\beta = (\beta_1, 1-\beta_1) \in \mathbb{R}^2$ based on a first order approximation of the minimal value function. Let the Assumption 4.12 hold. We assume the matrix Φ, the point f^*, the direction v, and the normal unit vector $\bar{n}$ to be given.

Note, that the normal boundary intersection method corresponds to the modified Pascoletti-Serafini problem with the constraint

$$a + t\, r - f(x) = 0_2.$$

Thus we always have $k = a + t\, r - f(x) = 0_2$. Moreover, the approximation points gained with this algorithm are not necessarily weakly EP-minimal points. In Chap. 5 this algorithm is applied to a convex test problem where all obtained approximation points are EP-minimal.

Algorithm 3 (Adaptive NBI method).
- **Input:** Choose the desired distance $\alpha \in \mathbb{R}$, $\alpha > 0$, between the approximation points.
- **Step 1:** Set $\beta^1 := (1,0)^\top$ and solve (NBI(β)) with solution $(\bar{s}^1, \bar{x}^1) = (0, x^1)$ and Lagrange multiplier μ^1 to the constraint $f^* + \Phi \beta^1 + s\, \bar{n} - f(x) = 0_2$. Set $l := 1$.
- **Step 2:** Set $\beta^E := (0,1)^\top$ and $v := \Phi \begin{pmatrix} -1 \\ 1 \end{pmatrix}$.
- **Step 3:** Set

$$\beta^{l+1} := \beta^l + \frac{\alpha}{\|v + ((\mu^l)^\top v)\bar{n}\|} \cdot \begin{pmatrix} -1 \\ 1 \end{pmatrix}$$

and $l := l + 1$.
- **Step 4:** If $\beta_1^l \geq 0$ solve (NBI(β)) with solution (s^l, x^l) and Lagrange multiplier μ^l to the constraint $f^* + \Phi \beta^l + s\, \bar{n} - f(x) = 0_2$ and go to Step 3. Else stop.
- **Step 5:** Determine the set $\tilde{A} := \{x^1, \ldots, x^{l-1}, x^E\}$ and the set $A := \mathcal{M}(\tilde{A}, \mathbb{R}_+^2)$ of non-dominated points of $\tilde{A}$.
- **Output:** The set A is an approximation of the set $\mathcal{M}(f(\Omega), \mathbb{R}_+^2)$.

Note that the points of the set $\tilde{A}$ are not necessarily weakly EP-minimal.

4.2.4 Algorithm for the Modified Polak Scalarization

The modified Polak method is very similar to the ε-constraint method and so is the algorithm. The scalar optimization problem ($MP(y_1)$) is

4.2 Parameter Control in the Bicriteria Case

$$\min\ f_2(x)$$
subject to the constraints
$$f_1(x) = y_1,$$
$$x \in \Omega.$$

This problem can again be considered as a special case of the modified Pascoletti-Serafini scalarization $(\overline{SP}(a,r))$. Here it is (compare Lemma 2.33) $H = \{y \in \mathbb{R}^2 \mid y_2 = 0\}$, i.e. $b = (0,1)^\top$, $\beta = 0$, and $r = (0,1)^\top$. The algorithm for this special case reads as follows, again assuming the Assumption 4.12 to be satisfied.

Algorithm 4 (Adaptive modified Polak method).

Input: Choose the desired distance $\alpha \in \mathbb{R}$, $\alpha > 0$, between the approximation points.

Step 1: Determine the numbers $y_1^1 := f_1(x^1) := \min_{x \in \Omega} f_1(x)$ and $y_1^E := f_1(x^E)$ with $f_2(x^E) := \min_{x \in \Omega} f_2(x)$.

Step 2: Solve $(MP(y_1^1))$ with minimal solution x^1 and Lagrange multiplier $\mu^1 \in \mathbb{R}$ to the constraint $y_1^1 - f_1(x) = 0$. Set $l := 1$.

Step 3: Set
$$y_1^{l+1} := y_1^l + \frac{\alpha}{\sqrt{1 + (\mu^l)^2}}$$
and $l := l + 1$.

Step 4: If $y_1^l \leq y_1^E$ solve $(MP(y_1^l))$ with minimal solution x^l and Lagrange multiplier μ^l to the constraint $y_1^l - f_1(x) = 0$ and go to Step 3. Else stop.

Step 5: Determine the set $\tilde{A} := \{x^1, \ldots, x^{l-1}, x^E\}$ and the set $A := \mathcal{M}(\tilde{A}, \mathbb{R}_+^2)$ of non-dominated points of $\tilde{A}$.

Output: The set A is an approximation of the set $\mathcal{M}(f(\Omega), \mathbb{R}_+^2)$.

An algorithm for the modified Polak method without an adaptive parameter control can be found in [124, Alg. 12.1]. Note, that the approximation points gained with this algorithm are not necessarily weakly EP-minimal. In Step 2 problem $(MP(y_1^1))$ can lead to numerical difficulties because it can happen that the constraint set is reduced to one point only. For avoiding this y_1^1 can be replaced by $y_1^1 + \Delta \varepsilon$ with a small value $\Delta \varepsilon > 0$.

4.3 Adaptive Parameter Control in the Multicriteria Case

In this section we discuss how to proceed if the multiobjective optimization problem (MOP) has three or more objective functions. We first recall the reasons why the results for the bicriteria case cannot just be generalized. Then we adapt our method by determining the approximation of the efficient set in two steps. In the first step we calculate a simple coarse approximation of the efficient set without using special results on an adaptive parameter control. This set is a base for detecting the interesting parts of the efficient set. For example the parts where the efficient set shows a particular behavior or the parts, where the most preferred solutions (by the d. m.) of all efficient points are supposed. In these special parts the approximation of the efficient set is now refined in the second step. This refinement is done by using sensitivity information for controlling the quality of the refinement.

As mentioned the parameter control for multiobjective optimization problems for three or more objective functions is, compared to the parameter control in the bicriteria case, a much more difficult task as different problems occur (compare [139]). First, in contrast to the bicriteria case with a two-dimensional image space (see Lemma 1.20), not any closed pointed convex cone in $\mathbb{R}^m$ ($m \geq 3$) is polyhedral. For example the Löwner partial ordering cone in $\mathbb{R}^3$ given in Example 1.21 is not polyhedral.

We have also problems in higher dimensions in the case that the ordering cone is polyhedral as we have already discussed in Sect. 2.3.2. There we have also already determined a set H^0 so that it is sufficient to consider parameters $a \in H^0$, compare Lemma 2.20. In contrast to the bicriteria case, where it is sufficient to consider a line segment H^a for the parameter a, the set H^0 is generally no longer (a part of) a line. So, it is an additional task to order the parameters a on this hyperplane. The order is important because we want to use sensitivity information, i.e. information about the local behavior of the local minimal value function, for determining nearby parameters a. For doing this we need the information which of the already found approximation points lie in the neighborhood of the next point, which we plan to determine. Getting this information can be time-consuming especially in higher dimensions.

For avoiding these problems we start with a coarse equidistant discretization of the set H^0 which delivers the first parameters a. This

4.3 Parameter Control in the Multicriteria Case

discretization of the $(m-1)$-dimensional cuboid H^0 is done with a fineness defined by the d.m.. The hyperplane including the set H^0 is spanned by the vectors v^i, $i = 1, \ldots, m-1$, see Sect. 2.3.2. The d.m. can for instance set the number $N^i \in \mathbb{N}$ of discretization points in each direction v^i, $i = 1, \ldots, m-1$. Then we get the following set of $\prod_{i=1}^{m-1} N^i$ equidistant discretization points:

$$D^{H^0} := \left\{ y \in H^0 \;\middle|\; y = \sum_{i=1}^{m-1} s_i v^i,\; s_i \in \{s_i^{\min,i} + \frac{L_i}{2} + l \cdot L_i \;\middle|\; l = 0, \ldots, N^i - 1\},\; i = 1, \ldots, m-1 \right\}$$

with

$$L_i := \frac{s_i^{\max,i} - s_i^{\min,i}}{N^i}.$$

For each of these discretization points $d \in D^{H^0}$ the scalar optimization problem (SP(d, r))

$$\min t$$

subject to the constraints

$$d + tr - f(x) \in K,$$
$$t \in \mathbb{R},\; x \in \Omega$$

has to be solved. Let (t^d, x^d) denote a minimal solution of this problem (if a solution exists). Then, the set

$$D^{H^0, f} := \{ f(x^d) \mid \exists t^d \in \mathbb{R} \text{ and a } d \in D^{H^0} \text{ with } (t^d, x^d) \text{ a minimal solution of (SP(d,r))} \}$$

determines a first approximation of the efficient set. Note, that generally the cardinality of the set D^{H^0} is higher then the cardinality of $D^{H^0, f}$, as different parameters d can result in the same approximation point $f(x^d)$.

The set $D^{H^0, f}$ is a coarse approximation of the efficient set based on which the d.m. can decide in which areas he or she is especially interested. The smaller the numbers N^i, $i = 1, \ldots, m-1$, are, the less points have to be interpreted but the more is the efficient set misrepresented.

As the following procedure is based on the fact that the d.m. chooses some interesting parts by selecting several points, it has an interactive component. In the next step the approximation is refined around the chosen points. For the refinement it is the aim to determine equidistant refinement points in the image space. For this aim sensitivity information is used (like in the bicriteria case). We describe the procedure for an arbitrary chosen point

$$\bar{y} := y^d := f(x^d) \in D^{H^0,f}.$$

Thus (t^d, x^d) is a minimal solution of (SP(d,r)) for a parameter $d \in D^{H^0}$. We determine the parameter $\bar{a}$ by

$$\bar{a} := f(x^d) - \bar{t}r \quad \text{with} \quad \bar{t} := \frac{b^\top f(x^d) - \beta}{b^\top r}.$$

Thus, the inequality constraint $\bar{a} + tr - f(x) \in K$ is active in the point $(\bar{t}, x^d)$. As we have already solved the problem (SP(d,r)) we have thereby gained information about the local behavior of the local minimal value function.

In the following we presume that the assumptions of Lemma 3.14 are satisfied, i.e. (among others) we presume $S = \mathbb{R}^n$, $K = \mathbb{R}^m_+$, $C = \mathbb{R}^p_+$. Then we have $\nabla_a \tau^\delta(\bar{a}, r) = -\bar{\mu}$ with $(\bar{\mu}, \bar{\nu}, \bar{\xi})$ the Lagrange multipliers to the point $(\bar{t}, x^d)$ for the problem (SP($\bar{a},r$)). Note that these are equal to the Lagrange multipliers to the point (t^d, x^d) for the optimization problem (SP(d,r)), compare Lemma 2.26. We get the approximation

$$t(a,r) \approx \bar{t} - \bar{\mu}^\top (a - \bar{a}) \tag{4.27}$$

for the minimal value $t(a,r)$ of the optimization problem (SP(a,r)) for a in a neighborhood of $\bar{a}$ (compare (4.6)).

If the d.m. is interested in a distance of $\alpha > 0$ between the approximation points of the refinement around the point y^d, a new approximation point with distance α can be determined in each direction v^i, $i = 1, \ldots, m-1$. For instance if a new approximation point should be found in direction v^k ($k \in \{1, \ldots, m-1\}$), i.e. $a^k = \bar{a} + s \cdot v^k$, we get with (4.27) and $f(\bar{x}) = \bar{a} + \bar{t}r$

$$\begin{aligned} f(x(a^k)) &= a^k + t(a^k, r)\, r \\ &\approx \bar{a} + s\, v^k + \bar{t}r - \bar{\mu}^\top (a^k - \bar{a})\, r \\ &= f(x^d) + s\left(v^k - (\bar{\mu}^\top v^k) r\right) \\ &= \bar{y} + s\left(v^k - (\bar{\mu}^\top v^k) r\right). \end{aligned}$$

Thus, the aim $\|f(x(a^k)) - \bar{y}\| = \alpha$ is approximately satisfied for

4.3 Parameter Control in the Multicriteria Case

$$|s| = \frac{\alpha}{\|v^k - (\bar{\mu}^\top v^k)r\|}.$$

We set $s^k := |s|$ and get the two new parameters

$$a^{k,1} = \bar{a} + s^k v^k \text{ and } a^{k,2} = \bar{a} - s^k v^k.$$

These steps can be repeated for all $k \in \{1, \ldots, m-1\}$. Then, for all parameters $a^{k,i}$, $k = 1, \ldots, m-1$, $i = 1, 2$, the scalar-valued optimization problem (SP($a^{k,i}, r$)) has to be solved. Let $(t^{k,i}, x^{k,i})$ denote a minimal solution, then for the point $f(x^{k,i})$ it is $\|f(x^{k,i}) - \bar{y}\| \approx \alpha$, with the accuracy dependent on the accuracy of the used sensitivity results.

The described procedure can be modified as follows: the d.m. determines one direction $v \in H$ in which additional information about the efficient set is desired. The additional parameters a can be calculated analogously. Besides, the determination of further approximation points with distances $2\alpha, 3\alpha, 4\alpha, \ldots$ is possible.

By solving the scalar optimization problems (SP($a^{k,i}, r$)) we get again sensitivity information for the local minimal value function around the point $a^{k,i}$. Thus, the points $f(x^{k,i})$, $k = 1, \ldots, m-1$, $i = 1, 2$, can be added to the set $D^{H^0, f}$ and the whole procedure (i.e. choice of the d.m., determination of refinement points) can be repeated arbitrarily.

We summarize the described steps in an algorithm. We assume $K = \mathbb{R}^3_+$, $C = \mathbb{R}^p_+$, $\hat{S} = S = \mathbb{R}^n$, and further let Assumption 4.12 hold, here for $m = 3$. We give the algorithm for simplicity for tricriteria optimization problems (i.e. $m = 3$):

$$\min \begin{pmatrix} f_1(x) \\ f_2(x) \\ f_3(x) \end{pmatrix}$$

subject to the constraints

$$g(x) \leq 0_p,$$
$$h(x) = 0_q,$$
$$x \in \mathbb{R}^n.$$

Further, we use a special case of the Pascoletti-Serafini scalarization. We assume $b = (0, 0, 1)^\top$, $\beta = 0$, $H = \{y \in \mathbb{R}^3 \mid y_3 = 0\}$, and $r = (0, 0, 1)^\top$. Thus, the problem (SP(a, r)) is for $a = (\varepsilon_1, \varepsilon_2, 0)$ equivalent to the ε-constraint problem ($P_3(\varepsilon)$)

4 Adaptive Parameter Control

$$\min f_3(x)$$
$$\text{subject to the constraints}$$
$$\varepsilon_1 - f_1(x) \geq 0,$$
$$\varepsilon_2 - f_2(x) \geq 0,$$
$$x \in \Omega$$

for $\varepsilon = (\varepsilon_1, \varepsilon_2) \in \mathbb{R}^2$ (compare Theorem 2.27).

Algorithm 5 (Adaptive ε-constraint method for tricriteria problems).

Input: Choose the desired number N_1 of discretization points for the range of the function f_1 (i.e. in direction $v^1 := (1,0,0)^\top$) and N_2 for the range of the function f_2 (i.e. in direction $v^2 := (0,1,0)^\top$).

Step 1: Solve the optimization problems $\min_{x \in \Omega} f_i(x)$ with minimal solution $x^{\min,i}$ and minimal value $f_i(x^{\min,i}) =: \varepsilon_i^{\min}$ for $i = 1, 2$ as well as $\max_{x \in \Omega} f_i(x)$ with maximal solution $x^{\max,i}$ and maximal value $f_i(x^{\max,i}) =: \varepsilon_i^{\max}$ for $i = 1, 2$.

Step 2: Set
$$L_i := \frac{\varepsilon_i^{\max} - \varepsilon_i^{\min}}{N_i} \quad \text{for } i = 1, 2$$

and solve the problem $(P_3(\varepsilon))$ for all parameters $\varepsilon \in E$ with

$$E := \left\{ \varepsilon = (\varepsilon_1, \varepsilon_2) \in \mathbb{R}^2 \,\middle|\, \varepsilon_i = \varepsilon_i^{\min} + \frac{L_i}{2} + l_i \cdot L_i \right.$$
$$\left. \text{for } l_i = 0, \ldots, N_i - 1, \ i = 1, 2 \right\}.$$

Determine the set

$$A^E := \left\{ (\varepsilon, \bar{x}, \bar{\mu}) \,\middle|\, \bar{x} \text{ is a minimal solution of } (P_3(\varepsilon)) \text{ with} \right.$$
$$\text{parameter } \varepsilon \text{ and Lagrange-multiplier } \bar{\mu}$$
$$\text{to the constraints } f_i(x) \leq \varepsilon_i, i = 1, 2$$
$$\left. \text{for } \varepsilon \in E \right\}.$$

Step 3: Determine the set

$$D^{H^0,f} := \left\{ f(x) \,\middle|\, \exists \varepsilon \in \mathbb{R}^2,\ \mu \in \mathbb{R}^2_+ \text{ with } (\varepsilon, x, \mu) \in A^E \right\}.$$

Input: Choose $y \in D^{H^0,f}$ with $y = f(x^\varepsilon)$ and $(\varepsilon, x^\varepsilon, \mu^\varepsilon) \in A^E$.
If y is a sufficient good solution, then stop.
Else, if additional points in the neighborhood of y are desired, give a distance $\alpha \in \mathbb{R}$, $\alpha > 0$, in the image space and the number of desired new points $\bar{n} = (2k+1)^2 - 1$ (for a $k \in \mathbb{N}$) and go to Step 4.

Step 4: Set

$$\varepsilon^{i,j} := \varepsilon + i \cdot \frac{\alpha}{\sqrt{1+(\mu_1^\varepsilon)^2}} \begin{pmatrix} 1 \\ 0 \end{pmatrix} + j \cdot \frac{\alpha}{\sqrt{1+(\mu_2^\varepsilon)^2}} \begin{pmatrix} 0 \\ 1 \end{pmatrix}$$

for all

$$(i,j) \in \left\{ (i,j) \in \mathbb{Z}^2 \,\middle|\, i, j \in \{-k, \ldots, k\},\ (i,j) \neq (0,0) \right\}$$

and solve problem $(P_3(\varepsilon^{i,j}))$.

If there exists a solution $x^{i,j}$ with Lagrange multiplier $\mu^{i,j}$, then set $A^E := A^E \cup \{(\varepsilon^{i,j}, x^{i,j}, \mu^{i,j})\}$.
Go to Step 3.

Output: The set $D^{H^0,f}$ is an approximation of the set of weakly efficient points $\mathcal{E}_w(f(\Omega), \mathbb{R}^3_+)$.

By the coarse approximation done in Step 3 it is assured that no parts of the efficient set are neglected and that the d. m. gets a survey about the characteristics of the efficient set. In the following the method is partly interactive as the d. m. chooses interesting areas as well as the fineness of the refinement. Of course, more than one point $y \in D^{H^0,f}$ can be chosen in the second input step for an improvement of the approximation of the efficient set. In Step 4 we use first order sensitivity information for the parameter determination but approximations of higher-order can also be applied.

5
Numerical Results

For testing the numerical methods developed in Chap. 4 on their efficiency, in this chapter several test problems with various difficulties are solved. Test problems are important to get to know the properties of the discussed procedures. Large collections of test problems mainly developed for testing evolutionary algorithms are given in [41, 45, 46, 228]. Some of the problems of these collections test difficulties especially designed for evolutionary algorithms, which are not a task using the scalarization approaches of this book. For instance, arbitrarily generated points of the constraint set are mapped by the objective functions mainly in areas far away from the efficient set and only few of them have images near the efficient set.

Other test problems have a non-convex image set or gaps in the efficient set, i.e. non-connected efficient sets. We examine such problems among others in this chapter. In the remaining chapters of this book, we discuss the application of the methods of Chap. 4 on a concrete application problem in medical engineering as well as on a multiobjective bilevel optimization problem.

5.1 Bicriteria Test Problems

First we examine multiobjective optimization problems with two objective functions and we apply the methods of Sect. 4.2.

5.1.1 Test Problem 1: ε-Constraint Scalarization

We start with a simple problem by Hazen ([103, p.186]) which we have already considered in Example 4.7. The bicriteria optimization problem

w. r. t. the natural ordering is

$$\min_{x \in \mathbb{R}^2} \begin{pmatrix} \frac{x_1^2}{2} + x_2^2 - 10x_1 - 100 \\ x_1^2 + \frac{x_2^2}{2} - 10x_2 - 100 \end{pmatrix}.$$

It is an unconstraint problem with a convex image set. In the biobjective case it is guaranteed that the Pareto set is the graph of a convex function, if the objective functions and the constraint set are convex ([210, Theorem 12]).

We solve this bicriteria optimization problem using the ε-constraint scalarization ($P_2(\varepsilon)$). In the Steps 1 and 2 of the Algorithm 2 as presented in Sect. 4.2.2 the minimal solutions x^1 and x^E are determined. We get that only parameters ε with

$$\varepsilon \in [f_1(x^E), f_1(x^1)] = [-150, 0]$$

have to be considered.

If we approximate the efficient set of this multiobjective optimization problem with the help of the scalarization ($P_2(\varepsilon)$) without an adaptive parameter control just by choosing equidistant parameters ε in the interval $[-150, 0]$, we get the approximation of the efficient set shown in Fig. 5.1. The parameters ε are also plotted in the figure as points $(\varepsilon, 10) \in \mathbb{R}^2$ marked with crosses. Here, a distance of 8 has been chosen between the parameters ε.

This approximation is much improved using Alg. 2 of Sect. 4.2.2 with the adaptive parameter control. Here, we predefine the distance $\alpha = 12$ (between the approximation points in the image space). We get again 20 approximation points but now evenly spread, as it can be seen in Fig. 5.2. This approximation can be improved using second order sensitivity information as described in Sect. 4.2.2. This results in the 21 approximation points drawn in Fig. 5.3.

For a comparison the quality criteria as discussed in Sect. 4.1 are evaluated for the three approximations. For the number N of approximation points, the rounded values of the uniformity level δ and of the approximated coverage error $\bar{\varepsilon}$ see Table 5.1. As we predefine $\alpha = 12$ the aim is $\bar{\varepsilon} = 6$ and $\delta = 12$.

Note, that we have $\delta = 11.1532$ and $\delta = 10.5405$ for the cases two and three because at the end of the applied algorithms the point $f(x^E)$ is added to the approximation point set. However thereby no distance control takes place. Excluding this point results in the uniformity $\delta = 12.0012$ for the Case 2 and $\delta = 12.011$ for the Case 3.

5.1 Bicriteria Test Problems   143

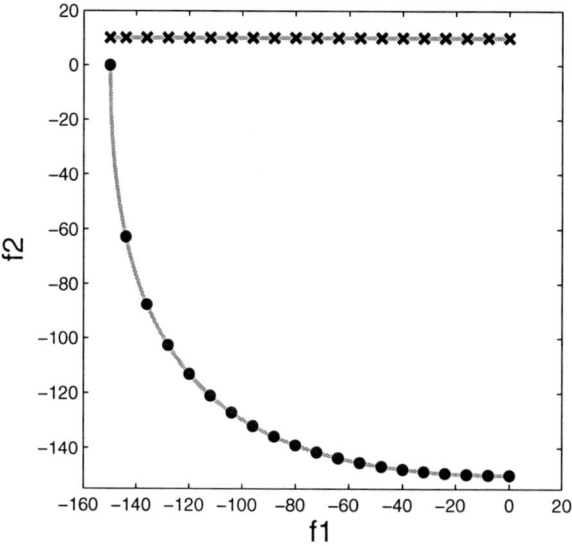

Fig. 5.1. Test problem 1: Approximation with equidistant parameters.

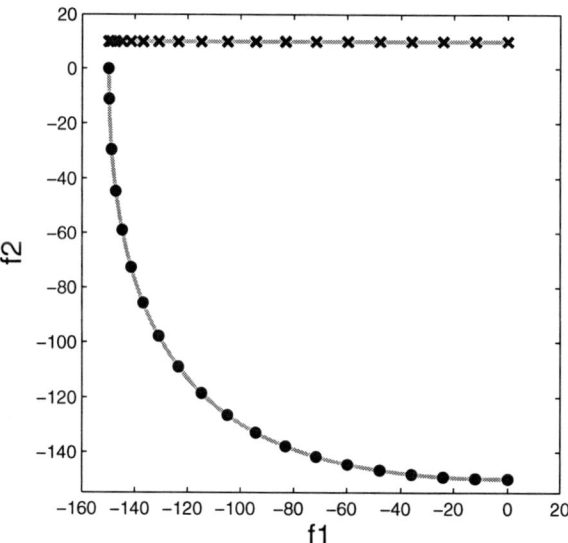

Fig. 5.2. Test problem 1: Approximation with the adaptive parameter control using first order sensitivity information according to Alg. 2.

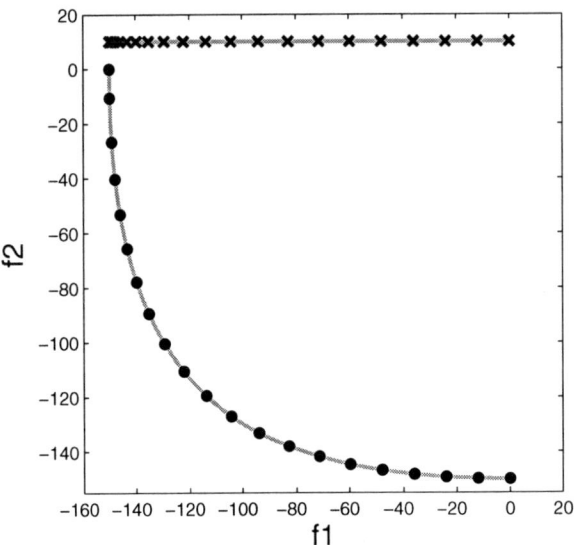

Fig. 5.3. Test problem 1: Approximation with the adaptive parameter control using second order sensitivity information.

Table 5.1. Quality criteria for the approximations of test problem 1.

	Case 1: equal parameter choice	Case 2: first order adaptive parameter choice	Case 3: second order adaptive parameter choice	aim
N	20	20	21	
$\bar{\varepsilon}$	31.5184	9.1963	8.0536	6
δ	8.003	11.1532	10.5405	12

5.1.2 Test Problem 2: Comparison with the Weighted Sum Method

Here, we compare the procedures of Chap. 4 with the wide-spread weighted sum scalarization. For a bicriteria optimization problem the weighted sum scalarization (see Sect. 2.5.7) is given by

$$\min_{x \in \Omega} w_1 f_1(x) + w_2 f_2(x)$$

with weights $w_1, w_2 \in [0,1]$, $w_1 + w_2 = 1$. By varying the parameters w_1 and w_2 and solving the weighted sum problem repeatedly approximations of the efficient set can be generated.

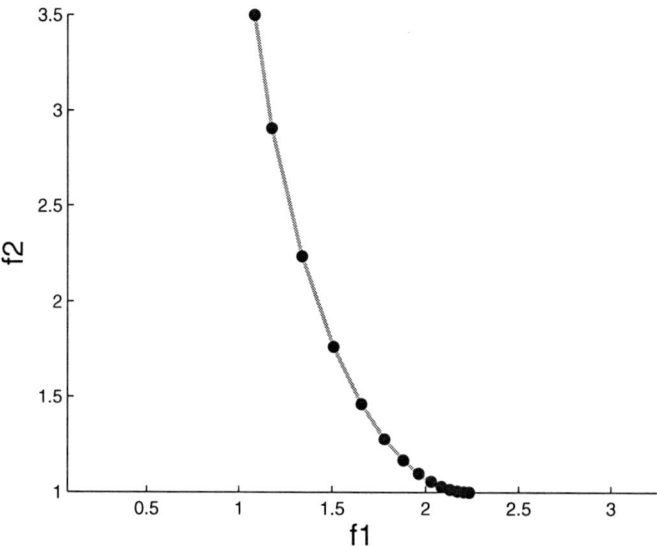

Fig. 5.4. Test problem 2: Approximation based on the weighted sum method.

For a comparison with the adaptive methods of this book we consider the following bicriteria optimization problem with a convex image set

$$\min \begin{pmatrix} \sqrt{1+x_1^2} \\ x_1^2 - 4x_1 + x_2 + 5 \end{pmatrix}$$

subject to the constraints

$$x_1^2 - 4x_1 + x_2 + 5 \leq 3.5,$$
$$x_1 \geq 0, \ x_2 \geq 0,$$
$$x \in \mathbb{R}^2$$

w.r.t. the natural ordering. For determining an approximation of the efficient set with 15 points we choose the 15 equidistant weights

$$\begin{pmatrix} w_1 \\ w_2 \end{pmatrix} \in \left\{ \begin{pmatrix} 0 \\ 1 \end{pmatrix}, \begin{pmatrix} 0.07 \\ 0.93 \end{pmatrix}, \ldots, \begin{pmatrix} 0.98 \\ 0.02 \end{pmatrix} \right\}.$$

This results in the non-uniform approximation shown in Fig. 5.4. Here, we have a visible low uniformity level and a high coverage error and thus not a high approximation quality.

We compare this approximation with an approximation generated with Alg. 1. We arbitrarily choose the hyperplane $H = \{y \in \mathbb{R}^2 \mid (1,1)y = 2.5\}$ and the direction $r = (1,0)^\top \in \mathbb{R}_+^2$. Further, we set $\alpha = 0.2$. Applying Alg. 1 results in the approximation with 15 points shown in Fig. 5.5. There, also the hyperplane H and the chosen parameters a (as crosses) are plotted.

Choosing $r = (0.1, 1)^\top$, $b = r$, $\beta = 1$, and still $\alpha = 0.2$ we get the approximation shown in Fig. 5.6,a) with 17 approximation points. This shows that the generated approximation is of course slightly influenced by the choice of the hyperplane H and the parameter r.

For completeness we apply also some of the algorithms for special cases of the (modified) Pascoletti-Serafini scalarization presented in Sect. 4.2. Starting with Alg. 2 for the ε-constraint scalarization results just in the special chosen parameters $b = (1,0)^\top$, $\beta = 0$, and $r = (1,0)^\top$. We get the approximation shown in Fig. 5.6,b). Here, the parameters ε are drawn as points $(0, \varepsilon)$ on the hyperplane $H = \{y \in \mathbb{R}^2 \mid y_1 = 0\}$.

Applying Alg. 3 for the normal boundary intersection scalarization results in the approximation plotted in Fig. 5.7,a). There, the hyper-

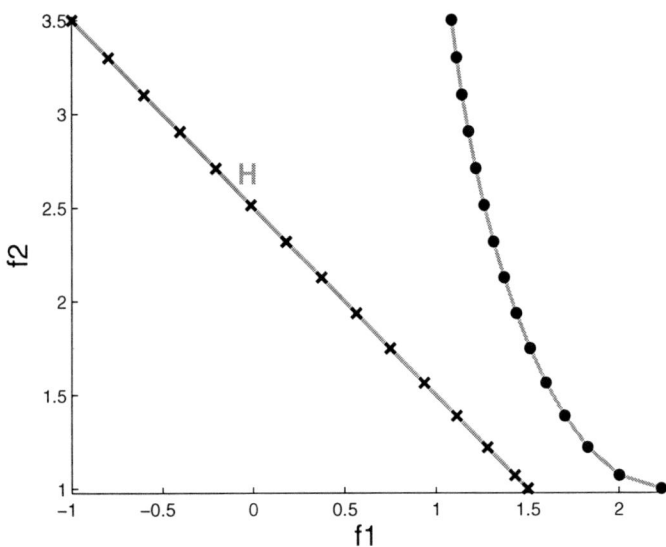

Fig. 5.5. Test problem 2: Approximation based on the adaptive Pascoletti-Serafini method.

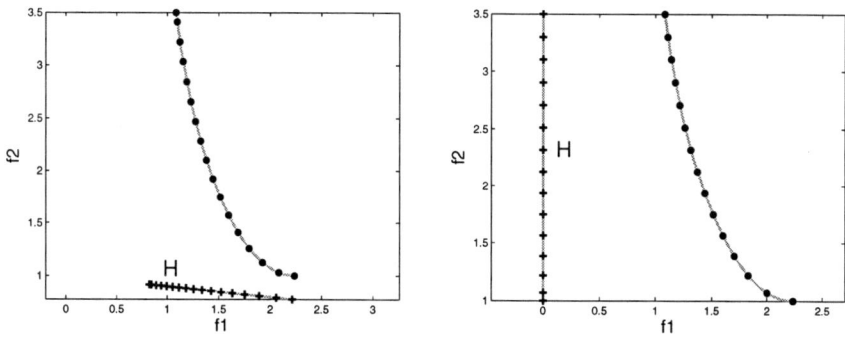

(a) with Alg. 1 and $r = (0.1, 1)^\top$, $b = r$, $\beta = 1$. (b) with Alg. 2 based on the ε-constraint scalarization.

Fig. 5.6. Test problem 2: Approximation of the efficient set

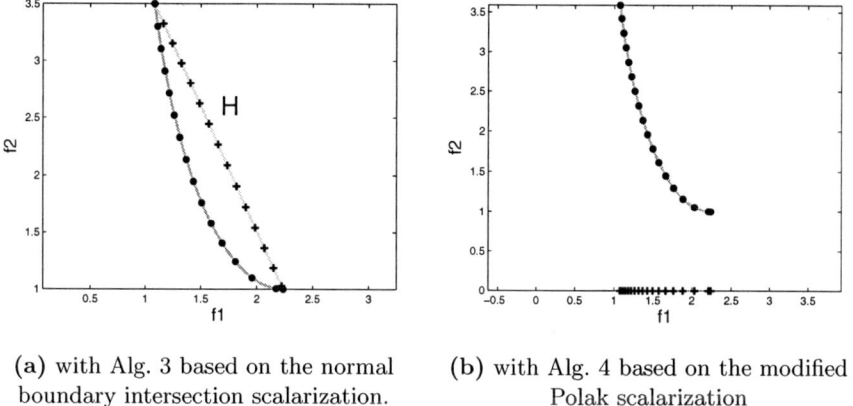

(a) with Alg. 3 based on the normal boundary intersection scalarization. (b) with Alg. 4 based on the modified Polak scalarization

Fig. 5.7. Test problem 2: Approximation of the efficient set

plane H is just the line connecting the two endpoints of the efficient curve. The parameter r is chosen as $-\bar{n}$, with $\bar{n}$ the normal unit vector of the hyperplane directing to the negative orthant.

Alg. 4 based on the modified Polak method corresponds to the hyperplane H with $b = (0, 1)^\top$, $\beta = 0$, and $r = (0, 1)^\top$. The generated approximation is given in Fig. 5.7,b).

5.1.3 Test Problem 3: Non-Convex Image Set

This test problem demonstrates the applicability of the numerical method summarized in Alg. 1 to non-convex problems. We consider the bicriteria optimization problem

$$\min \begin{pmatrix} 1 - \exp(-\sum_{i=1}^{n}(x_i - \frac{1}{\sqrt{n}})^2) \\ 1 - \exp(-\sum_{i=1}^{n}(x_i + \frac{1}{\sqrt{n}})^2) \end{pmatrix}$$

subject to the constraints

$$x_i \in [-4, 4], \quad i = 1, \ldots, n,$$

by van Veldhuizen ([228, p.545], also in [46, p.4]) with a non-convex image set. Let $K = \mathbb{R}_+^2$. Many scalarization problems have difficulties with non-convexity. For example using the weighted sum scalarization it is not possible to find all efficient points. Even by varying the weights arbitrarily only two efficient points can be found. These are the points $(0, 0.9817)$ and $(0.9817, 0)$.

Here, the dimension $n \in \mathbb{N}$ is a parameter which can be chosen arbitrarily. An interesting property of this test problem is the arbitrary scalability w.r.t. the parameter space dimension $n \in \mathbb{N}$ while the set of EP-minimal points is known explicitly. It is

$$\mathcal{M}(f(\Omega), \mathbb{R}_+^2) = \{x \in \mathbb{R}^n \mid x_1 \in [-\frac{1}{\sqrt{n}}, \frac{1}{\sqrt{n}}], \ x_i = x_1, \ i = 2, \ldots, n\}.$$

The efficient set is independently on the parameter n given by

$$\mathcal{E}(f(\Omega), \mathbb{R}_+^2) = \left\{ \begin{pmatrix} 1 - \exp(-4(t-1)^2) \\ 1 - \exp(-4t^2) \end{pmatrix} \bigg| \ t \in [0,1] \right\}.$$

Thus it is possible to determine the coverage error ε according to Sayin (see (4.2)), and not only an approximation of it, for any approximation A of the efficient set $\mathcal{E}(f(\Omega), \mathbb{R}_+^2)$. It is

$$\varepsilon = \max_{y \in \mathcal{E}(f(\Omega), K)} \min_{\bar{y} \in A} \|y - \bar{y}\|_2.$$

Hence, for an approximation $A = \{f(x^i) \mid i = 1, \ldots, N\}$, we have to solve (compare (4.3))

$$\max \delta$$

subject to the constraints

$$\delta \leq \left\| \begin{pmatrix} 1 - \exp(-4(t-1)^2) \\ 1 - \exp(-4t^2) \end{pmatrix} - f(x^i) \right\|_2, \quad i = 1, \ldots, N,$$

$$t \in [0, 1],$$

$$\delta \in \mathbb{R}.$$

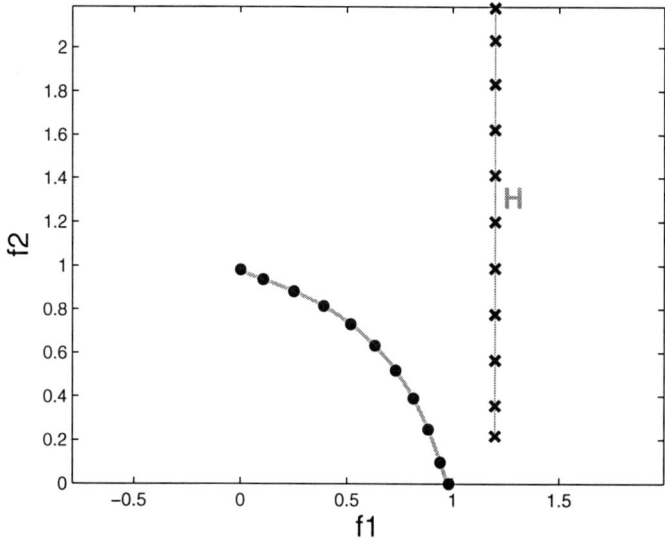

Fig. 5.8. Test problem 3: Non-convex bicriteria optimization problem.

We have determined an approximation of the efficient set of this test problem for $n = 40$ using Alg. 1 with $r = (1,1)^\top$, $b = (1,0)^\top$, $\beta = 1.2$, and for various distances α. For $\alpha = 0.15$ the generated approximation together with the hyperplane and the chosen parameters is shown in Fig. 5.8. Here, we have the coverage error $\varepsilon = 0.0790$, 15 approximation points (i.e. $N = 15$) and the uniformity level $\delta = 0.1058$. Ignoring the approximation point $f(x^E)$ which is added in the final step of Alg. 1 (and which is determined without parameter control) we get the corrected value $\delta^c = 0.1147$ for the uniformity level.

These quality criteria for the approximations gained with Alg. 1 for several distances of α are given in Table 5.2. There, also the values ε/α are given. Here, the aim is a value of $\varepsilon/\alpha = 0.5$. Further the corrected values for the uniformity level δ^c are tabulated, too.

5.1.4 Test Problem 4: Non-Connected Efficient Set

In test problem 3 the image set $f(\Omega)$ is non-convex but at least the efficient set is connected, i.e. there are no gaps. For the definition of a connected set (in the topological sense) see p. 104 (or [60, p.69], [190, p.66]).

Table 5.2. Quality criteria for the approximations of test problem 3 for different values of α.

	$\alpha = 0.05$	$\alpha = 0.1$	$\alpha = 0.15$	$\alpha = 0.2$	$\alpha = 0.25$	$\alpha = 0.3$	$\alpha = 0.4$
ε	0.0264	0.0530	0.0790	0.1046	0.1311	0.1546	0.2038
δ	0.0333	0.0420	0.1058	0.0645	0.1928	0.2314	0.3075
δ^c	0.0367	0.0755	0.1147	0.1538	0.1928	0.2314	0.3075
N	30	16	11	9	7	6	5
$\frac{\varepsilon}{\alpha}$	0.5273	0.5296	0.5264	0.5228	0.5245	0.5154	0.5094

Here, in this test problem by Tanaka ([221]) the image set $f(\Omega)$ is not only non-convex but the efficient set is also non-connected. The bicriteria problem w. r. t. the natural ordering is given by

$$\min \begin{pmatrix} x_1 \\ x_2 \end{pmatrix}$$

subject to the constraints

$$x_1^2 + x_2^2 - 1 - 0.1 \cos\left(16 \arctan(\tfrac{x_1}{x_2})\right) \geq 0,$$
$$(x_1 - 0.5)^2 + (x_2 - 0.5)^2 \leq 0.5,$$
$$x_1, x_2 \in [0, \pi],$$
$$x \in \mathbb{R}^2,$$

i. e. it is $f(\Omega) = \Omega$.

The EP-minimal points lie on the boundary of the set given by the first constraint. As this inequality constraint includes periodic functions (and because the second constraint has also to be satisfied), not all points of the boundary of this set are non-dominated. Thus the efficient set is non-connected. We use Alg. 1 with an arbitrarily chosen hyperplane given by $b = (1, 1)^\top$ and $\beta = 0.5$, the direction $r = (1, 2)^\top$, and we predefine the distance $\alpha = 0.08$.

Using only a local numerical solver for the scalar optimization problems appearing in Alg. 1 leads to difficulties, as the scalar problems (SP(a, r)) have several local minimal solutions besides the searched global minimal solution. With a numerical solver which is only able to detect local minimal solutions, it can happen that one generates only approximations like the one shown in Fig. 5.9. In Fig. 5.9 the boundary of the image set $f(\Omega)$ is drawn as well as the found efficient points (dots) and the chosen parameters on the hyperplane (crosses).

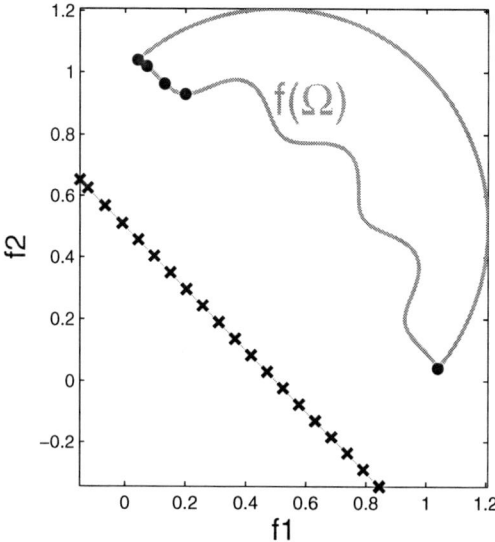

Fig. 5.9. Test problem 4: Non-connected efficient set approximated using a local scalar solution method.

Here, we have used the SQP-method (sequential quadratic programming method) as implemented in Matlab. As starting point for this method we used the minimal solution found for the previous parameter. The low approximation quality is not caused by the used adaptive parameter control but by the difficulty of finding global and not only local minimal solutions of a scalar optimization problem. We started with the parameter $a^1 = (-0.1517, 0.6517)$, but for the parameters $a^4 = (-0.0095, 0.5095)$ to $a^{20} = (0.8438, -0.3438)$ we got stuck in the EP-minimal point $x^4 = (0.1996, 0.9290)$.

Using a global solution method for the scalar optimization problems we get the approximation shown in Fig. 5.10. We see that the proposed adaptive parameter control can also handle non-connected efficient sets, but of course an appropriate global solution method for the scalar problems has to be applied. The need of global solvers for non-convex problems is a well-known fact for scalarization approaches.

Evaluating the quality criteria as discussed in Sect. 4.1 is easy for the cardinality and the uniformity level. For the approximation shown in Fig. 5.10 with 21 parameters we have 17 different non-dominated

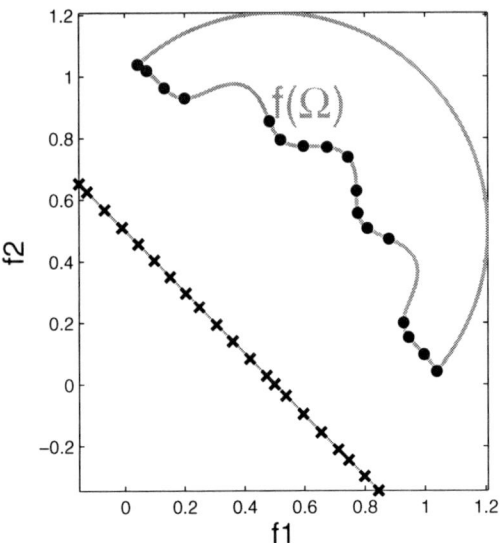

Fig. 5.10. Test problem 4: Non-connected efficient set approximated using global optimization for the scalar problems.

approximation points, i. e. $N = 17$, and

$$\delta = \min_{\substack{x,y \in A \\ x \neq y}} \|x - y\|_2 = \|f(x^2) - f(x^1)\|_2 = 0.0360.$$

For calculating the coverage error we have to take the non-connectedness of the efficient set into account. We approximate the coverage error by (4.4). Thus the set of neighbors have to be determined for each approximation point and for that we have to know where the gaps in the efficient set are. Here, the approximation points $f(x^i)$ for $i = 1, 4, 8, 16, 18, 21$ have each one neighbor only. With the distances between the approximation points given in Table 5.3 this leads to

$$\bar{\varepsilon} = \frac{1}{2} \max_{j \in \{1,\ldots,21\}} \max_{y \in \mathcal{N}(f(x^j))} \|f(x^j) - y\| = \frac{1}{2} \cdot 0.1133 = 0.0567$$

(here the aim is $\alpha/2 = 0.04$) with $\mathcal{N}(f(x^j))$ the set of neighbor points to the approximation point $f(x^j)$. The large distance between the points $f(x^7)$ and $f(x^8)$ and between the points $f(x^{17})$ and $f(x^{18})$ respectively is due to the gaps in the efficient set.

5.1 Bicriteria Test Problems

Table 5.3. Distances between consecutive approximation points to test problem 4.

j	1	2	3	4	5
$\|f(x^j) - f(x^{j+1})\|_2$	0.0360	0.0817	0.0762	0	0
j	6	7	8	9	10
$\|f(x^j) - f(x^{j+1})\|_2$	0	0.2928	0.0714	0.0792	0.0793
j	11	12	13	14	15
$\|f(x^j) - f(x^{j+1})\|_2$	0.0758	0.1133	0.0730	0.0585	0.0789
j	16	17	18	19	20
$\|f(x^j) - f(x^{j+1})\|_2$	0	0.2770	0.0502	0.0752	0.0695

5.1.5 Test Problem 5: Various Ordering Cones

For the examination of the effect of different partial orderings in the image space we consider the simple bicriteria optimization problem

$$\min \begin{pmatrix} x_1 \\ x_2 \end{pmatrix}$$

subject to the constraints (5.1)

$$(x_1 - 5)^2 + 4(x_2 - 3)^2 \leq 16,$$
$$x \in \mathbb{R}^2$$

w.r.t. various ordering cones. We calculate the set of K^0-minimal points for $K^0 := \mathbb{R}^2_+$, i.e. the set of EP-minimal points, as well as the set of $K^{\max}$-minimal points with $K^{\max} := -\mathbb{R}^2_+$, which is equivalent to determining the set of EP-maximal points. Further, we use the partial orderings given by the convex cones

$$K^1 := \{y \in \mathbb{R}^2 \mid (2, -1)y \geq 0, \ (-1, 2)y \geq 0\} \text{ and}$$
$$K^2 := \{y \in \mathbb{R}^2 \mid (0, 1)y \geq 0, \ (1, 1)y \geq 0\}$$

(see Fig. 5.11). These cones are defined by the matrices

$$\overline{K}^1 = \begin{pmatrix} 2 & -1 \\ -1 & 2 \end{pmatrix} \text{ and } \overline{K}^2 = \begin{pmatrix} 0 & 1 \\ 1 & 1 \end{pmatrix}$$

(compare p.15). The cone K^0 is generated by the unit matrix and $K^{\max}$ by the negative of the unit matrix.

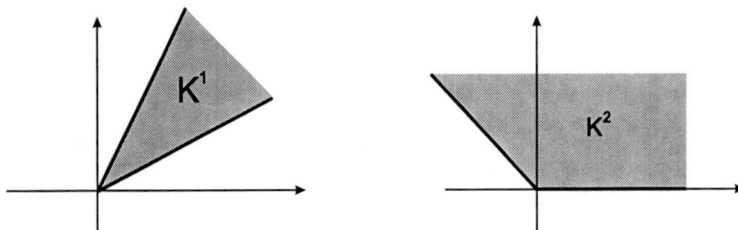

Fig. 5.11. Test problem 5: Ordering cones K^1 and K^2.

We have several possibilities to solve the bicriteria optimization problem w.r.t. these cones, as they are polyhedral cones. The scalarization (SP(a,r)) according to Pascoletti and Serafini for $K \in \{K^0, K^{\max}, K^1, K^2\}$ is

$$\min\ t$$
subject to the constraints
$$a + tr - f(x) \in K,$$
$$t \in \mathbb{R},\ x \in \Omega$$

with $\Omega := \{x \in \mathbb{R}^2 \mid (x_1 - 5)^2 + 4(x_2 - 3)^2 \leq 16\}$, being equivalent to

$$\min\ t$$
subject to the constraints
$$\overline{K}(a + tr - f(x)) \geq_2 0_2,$$
$$t \in \mathbb{R},\ x \in \Omega$$
(5.2)

for $\overline{K}$ the matrix which generates the cone K.

Another possibility for solving (5.1), if K is pointed, is the following reformulation (for the matrix $\overline{K}$ which corresponds to the cone K)

$$\min\ \overline{K} f(x)$$
subject to the constraint
$$x \in \Omega$$

with respect to the natural ordering (compare Lemma 1.18). The related scalarization according to Pascoletti and Serafini is then

$$\min\ t$$
subject to the constraints
$$a + tr - \overline{K} f(x) \geq_2 0_2,$$
$$t \in \mathbb{R},\ x \in \Omega.$$

However note, that using this reformulation we can use the described parameter control only for controlling the distances between the points $\overline{K}f(x^i)$ and not between the points $f(x^i)$. Thus for achieving the predefined distance between the points $f(x^i)$ further calculations involving the inverse of the matrix $\overline{K}$ are necessary.

Therefore we use the scalarization as given in (5.2). Here, we arbitrarily choose the hyperplane $H = \{y \in \mathbb{R}^2 \mid (1,1)y = 0\}$ and $r = (1,1)^\top \in K$ for $K \in \{K^0, K^1, K^2\}$ and $r = (-1,-1)^\top \in K$ for $K = K^{\max}$. Further we predefine $\alpha = 0.4$. The generated approximations for the several ordering cones are given in the Figures 5.12 and 5.13. As stated in Lemma 1.6 the larger the ordering cone the smaller is the efficient set. With this test problem it is shown that the proposed Alg. 1 also works for ordering cones which are not just representing the natural ordering.

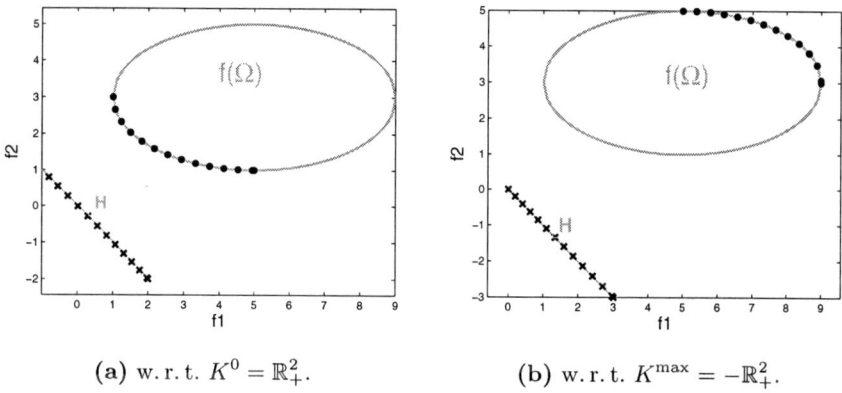

(a) w.r.t. $K^0 = \mathbb{R}_+^2$. (b) w.r.t. $K^{\max} = -\mathbb{R}_+^2$.

Fig. 5.12. Test problem 5: Approximation of the efficient set

5.2 Tricriteria Test Problems

This section is devoted to test problems with three objective functions. Here, we also discuss such difficulties as non-convexity or non-connectedness. We always apply Algorithm 5 to these test problems.

5.2.1 Test Problem 6: Convex Image Set

We start with a tricriteria optimization problem by Kim and de Weck ([139, p.8])

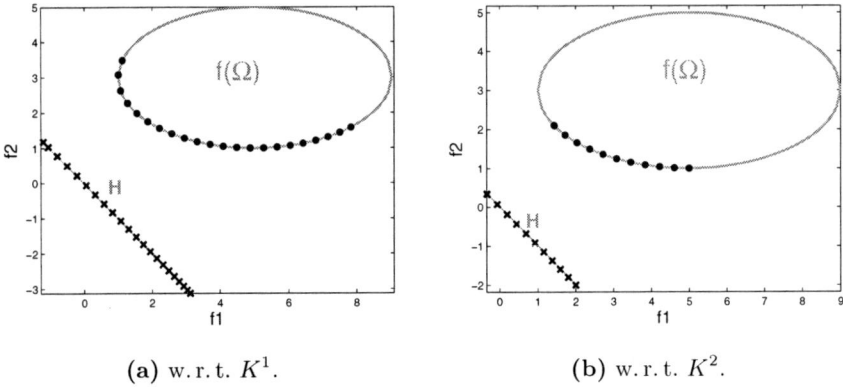

(a) w.r.t. K^1.   (b) w.r.t. K^2.

Fig. 5.13. Test problem 5: Approximation of the efficient set

$$\min \begin{pmatrix} -x_1 \\ -x_2 \\ -x_3 \end{pmatrix}$$

subject to the constraints
$$x_1^4 + 2x_2^3 + 5x_3^2 \leq 1,$$
$$x_i \geq 0, \quad i = 1, 2, 3,$$

w. r. t. the natural ordering. As the image set of this problem is convex the well-known weighted sum method is also able to find all efficient points. Thus, for the purpose of comparison we start with this wide spread method and so we have to solve

$$\min \ -(w_1 x_1 + w_2 x_2 + w_3 x_3)$$
subject to the constraint
$$x \in \Omega$$

with $\Omega := \{x \in \mathbb{R}^3_+ \mid x_1^4 + 2x_2^3 + 5x_3^2 \leq 1\}$ for weights $w_i \geq 0$, $i = 1, 2, 3$, $\sum_{i=1}^{3} w_i = 1$. Here, we choose the weights w_i as in [139] by

$$w_1 := \alpha_1 \alpha_2,$$
$$w_2 := (1 - \alpha_1)\alpha_2,$$
$$w_3 := 1 - \alpha_2$$

for $\alpha_1, \alpha_2 \in \{0, \frac{1}{5}, \frac{2}{5}, \ldots, 1\}$. This results in the approximation shown in Fig. 5.14. For achieving a better representation we have drawn the

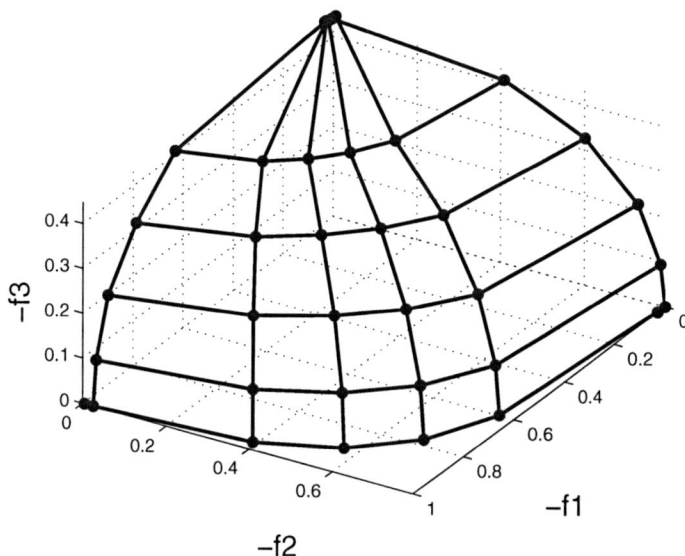

Fig. 5.14. Test problem 6: Approximation using the weighted sum scalarization.

negative of the objective function values. Besides we have connected the approximation points by lines. Due to the varying curvature of the efficient set some parts are well represented while other parts are neglected.

We compare this approximation with an approximation gained with Alg. 5. According to the algorithm we first have to solve the scalar problems

$$\min_{x \in \Omega} f_i(x) \text{ and } \max_{x \in \Omega} f_i(x) \text{ for } i = 1, 2,$$

resulting in the parameters $\varepsilon^{\min} = (-1, -0.7937)$ and $\varepsilon^{\max} = (0, 0)$. We choose $N^1 = N^2 = 5$. Then it is $L_1 = 0.2000$ and $L_2 = 0.1587$. For the parameters $\varepsilon = (\varepsilon_1, \varepsilon_2)$ with

$$\varepsilon_1 = -1 + \left(\frac{1}{2} + l_1\right) \cdot 0.2000 \text{ for } l_1 = 0, \ldots, 4,$$

$$\varepsilon_2 = -0.7937 + \left(\frac{1}{2} + l_2\right) \cdot 0.1587 \text{ for } l_2 = 0, \ldots, 4,$$

the ε-constraint problems ($P_3(\varepsilon)$) have to be solved:

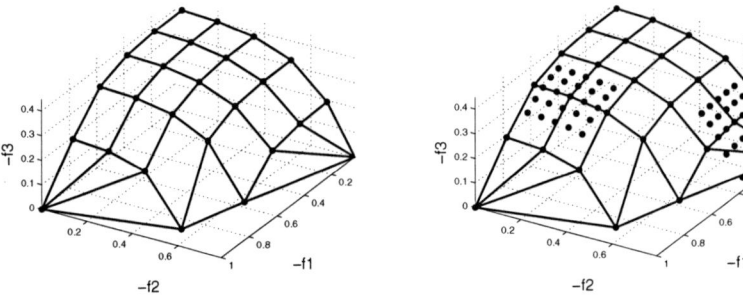

(a) with equidisdant parameter choice.

(b) with refinement in special chosen areas.

Fig. 5.15. Test problem 6: Approximation of the efficient set

$$\min -x_3$$
subject to the constraints
$$-x_1 \leq \varepsilon_1,$$
$$-x_2 \leq \varepsilon_2,$$
$$x \in \Omega.$$

The set of the images of the minimal solutions of the problems $(P_3(\varepsilon))$ for the several parameters ε under the function f, i.e. the set $D^{H^0,f}$, is shown in Fig. 5.15,a).

The approximation of Fig. 5.15,a) has a high coverage error, but it allows the d. m. to get a survey over the efficient set for selecting areas and approximation points respectively which are of interest. Based on this selection the approximation is now refined using the adaptive parameter control of Alg. 5 (Step 4). In this example we choose $k = 2$, $\alpha = 0.06$ and arbitrarily two approximation points and we get the refined approximation shown in Fig. 5.15,b).

5.2.2 Test Problem 7: Non-Convex Image Set

The following test problem (similar to an example in [138]) has a non-convex image set. We consider the tricriteria optimization problem w. r. t. the natural ordering:

5.2 Tricriteria Test Problems

$$\min \begin{pmatrix} -x_1 \\ -x_2 \\ -x_3^2 \end{pmatrix}$$

subject to the constraints

$$-\cos(x_1) - \exp(-x_2) + x_3 \leq 0,$$
$$0 \leq x_1 \leq \pi,$$
$$x_2 \geq 0,$$
$$x_3 \geq 1.2.$$

As this problem is non-convex the weighted sum scalarization is not an adequate method. Despite this, generating an approximation with that method – using the same procedure as described in Sect. 5.2.1 (test problem 6), now with parameters $\alpha_1, \alpha_2 \in \{0, \frac{1}{9}, \frac{2}{9}, \ldots, 1\}$ – results in the approximation shown in Fig. 5.16,a).

Here, we have again connected the approximation points by lines and we have drawn the negative of the objective function values. This approximation can even not be used for getting a coarse survey about the efficient set of the tricriteria optimization problem.

Using Alg. 5 leads first (Steps 1-3) to the approximation shown in Fig. 5.16,b). For applying the second part of the algorithm we choose $k = 2$ and $\alpha = 0.06$. Further, we assume that the d. m. realizes after evaluating the approximation of Fig. 5.16,b) that he is only interested

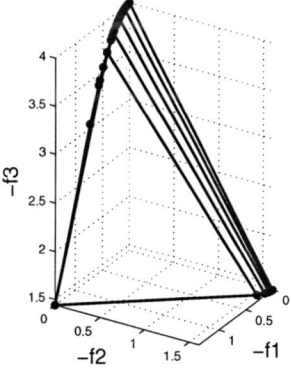
(a) using the weighted sum method.

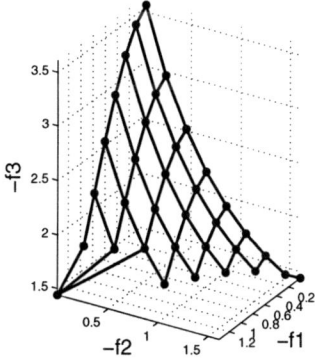
(b) with equidistant parameter choice using the ε-constraint scalarization.

Fig. 5.16. Test problem 7: Approximation of the efficient set

in the area of the efficient set for which it holds $f_1(x) \leq -0.4$ as well as $-0.6 \leq f_2(x) \leq -0.4$. According to Step 4 of Alg. 5 a refinement is done around the approximation points lying already in this area. The result is given in Fig. 5.17.

The correspondent parameters $\varepsilon = (\varepsilon_1, \varepsilon_2) \in \mathbb{R}^2$ for which we have solved the problem $(P_3(\varepsilon))$ in the course of Alg. 5 are drawn in Fig. 5.18. There, the parameters ε for which there exists no minimal solution of the problem $(P_3(\varepsilon))$ are marked with the smallest dots. With bigger dots the parameters are drawn which belong to the first coarse approximation. Finally, in gray, the refinement parameters are given. Here, one can see that the distances between the refinement parameters vary. This depends on the sensitivity information gained by solving the problem $(P_3(\varepsilon))$ to the parameter ε around which the refinement is done. The steeper the efficient set is in this area the smaller are the distances between the parameters.

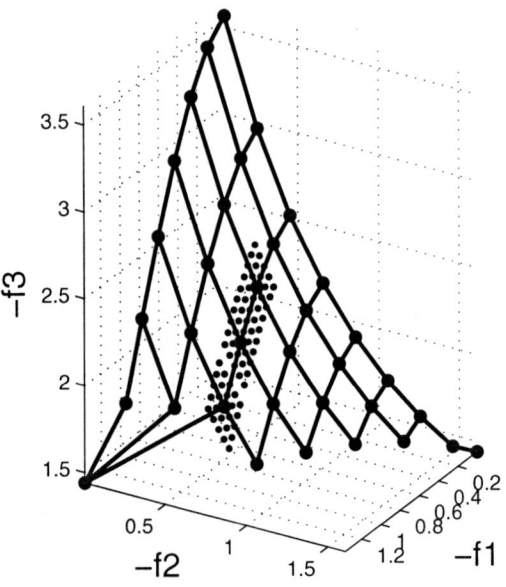

Fig. 5.17. Test problem 7: Refined approximation in special chosen areas.

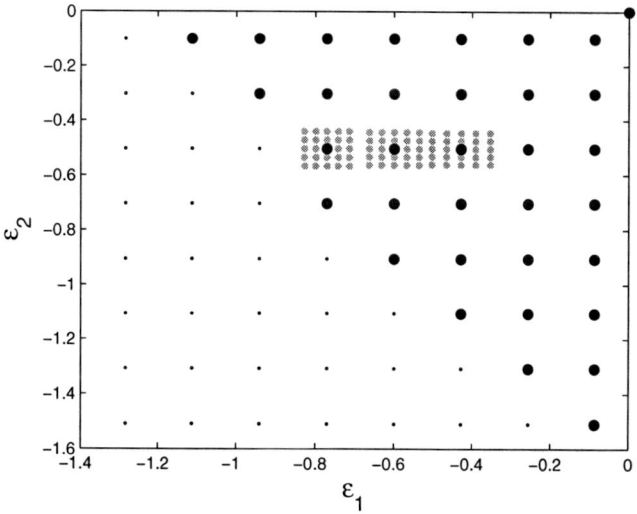

Fig. 5.18. Test problem 7: Parameter set.

5.2.3 Test Problem 8: Comet Problem

This problem taken from [46, p.9] carries its name due to the shape of the image set which looks like a comet with a small broad and a larger narrow part:

$$\min \begin{pmatrix} (1+x_3)(x_1^3 x_2^2 - 10x_1 - 4x_2) \\ (1+x_3)(x_1^3 x_2^2 - 10x_1 + 4x_2) \\ 3(1+x_3)x_1^2 \end{pmatrix}$$

subject to the constraints

$$1 \leq x_1 \leq 3.5,$$
$$-2 \leq x_2 \leq 2,$$
$$0 \leq x_3 \leq 1$$

w. r. t. the ordering cone $K = \mathbb{R}_+^3$. Here, the set of EP-minimal points is explicitly known and given by

$$\mathcal{M}(f(\Omega), \mathbb{R}_+^3) = \{x \in \mathbb{R}^3 \mid 1 \leq x_1 \leq 3.5, \ -2 \leq x_1^3 x_2 \leq 2, \ x_3 = 0\}.$$

The image set of this problem is shown in Fig. 5.19. For this representation we have discretized the whole feasible set with an equal

162   5 Numerical Results

distance of 0.1 and then we have mapped these points with the objective functions. The non-dominated points are drawn in gray. Note, that we plot again the negative objective function values.

Applying Alg. 5 covers up a weakness of this algorithm. In the first part (Steps 1-3) of the algorithm we solve for equidistantly chosen parameters $\varepsilon \in \mathbb{R}^2$ the ε-constraint problem $(P_3(\varepsilon))$. Here, we have to solve many of these problems for getting just very few approximation points of the efficient set. For example choosing $N^1 = N^2 = 12$ in Alg. 5, i.e. discretizing with 144 equidistant parameters, leads to only 7 approximation points in Step 3, shown in Fig. 5.20,a) (connected with lines).

Then we apply Step 4 of Alg. 5 with $k = 3$ and $\alpha = 4$. The additional approximation points gained by this refinement are added in Fig. 5.20,b). Thus, after this refinement step we have improved the quality of the approximation clearly observable.

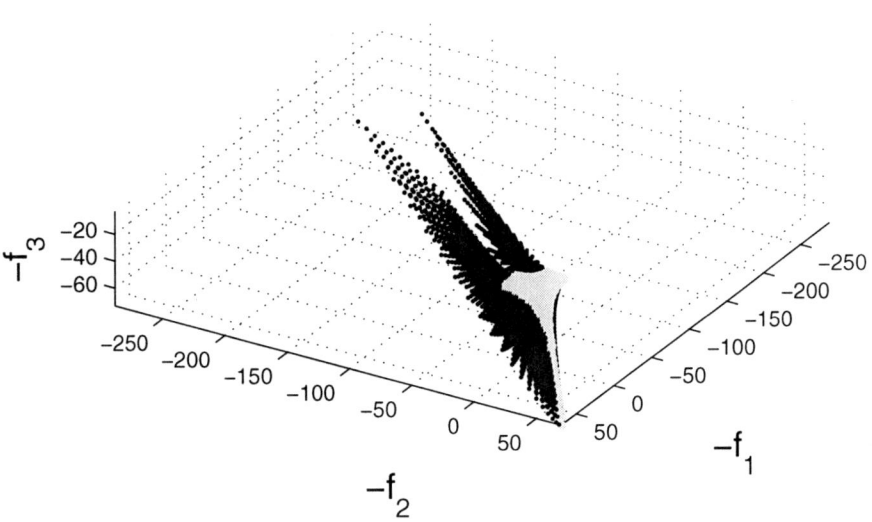

Fig. 5.19. Test problem 8: Image set and efficient set (in gray).

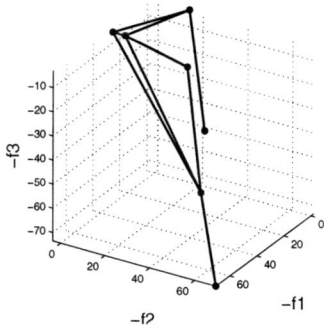
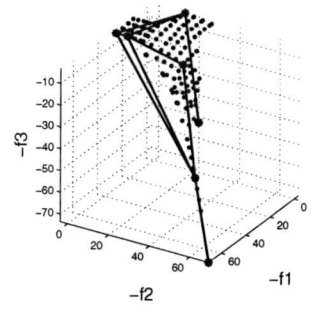

(a) with equidistant parameters.   (b) with refinement in special areas.

Fig. 5.20. Test problem 8: Approximation of the efficient set

Without an adaptive parameter control but with a usual procedure, the approximation is generally improved just by increasing the fineness of the equidistantly chosen parameters. Then, especially in this example, an above average number of additional scalar problems have to be solved for finding only comparatively few additional approximation points.

5.2.4 Test Problem 9: Non-Connected Efficient Set

The following test problem called DTLZ7 in [46] has not only a non-convex image set but even a non-connected efficient set:

$$\min \begin{pmatrix} x_1 \\ x_2 \\ (1+g(x))(3 - \sum_{i=1}^{2}\left(\frac{x_i}{1+g(x)}(1+\sin(3\pi x_i))\right)) \end{pmatrix}$$

subject to the constraints
$$x_i \in [0,1], \quad i = 1, \ldots, 22,$$

with

$$g(x) := 1 + \frac{9}{20}\sum_{i=3}^{22} x_i$$

and $K = \mathbb{R}^3_+$.

Here, the efficient set is separated in four non-connected parts. The set of EP-minimal points is a subset of the set

164   5 Numerical Results

$$\{x \in \mathbb{R}^{22} \mid x_i = 0, \ i = 3, \ldots, 22\},$$

and thus the efficient set is a subset of the set

$$B := \left\{ y \in \mathbb{R}^3 \,\middle|\, y_1, y_2 \in [0,1], \ y_3 = 2 \cdot \left(3 - \sum_{i=1}^{2} \left(\frac{y_i}{2}(1 + \sin(3\pi y_i)) \right) \right) \right\},$$

which is plotted in Fig. 5.21.

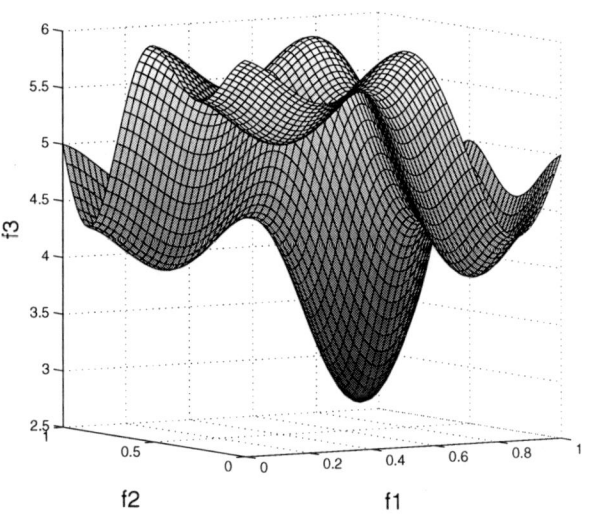

Fig. 5.21. Test problem 9: Set B.

We apply Alg. 5 with the following values: $N^1 = N^2 = 10$, $k = 2$, $\alpha = 0.15$ and for Step 4 we refine only around those approximation points y of the efficient set with $y \in [0, 0.3] \times [0, 0.3]$. The result is shown in Fig. 5.22. The four separated parts of the efficient set can easily be detected. In one of these parts the refinement with almost equidistant points in the image space is done.

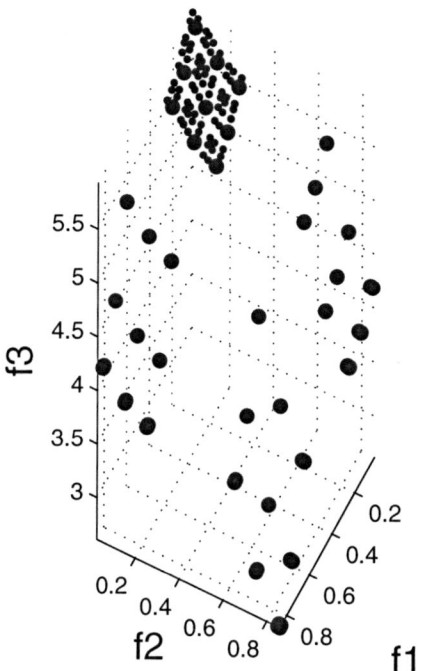

Fig. 5.22. Test problem 9: Approximation of the efficient set.

6

Application to Intensity Modulated Radiotherapy

As we have already pointed out in the introduction to this book, many problems arising in applications are from its structure multiobjective. Nevertheless these problems are often treated as a single objective optimization problem in practice. This was for instance done in intensity modulated radiotherapy (IMRT). Here, an optimal treatment plan for the irradiation of a cancer tumor has to be found for a patient. The aim is to destroy or at least reduce the tumor while protecting the surrounding healthy tissue from unnecessary damage. For a detailed problem description we refer to [3, 36, 63, 143, 146, 170, 223].

Regarding the natural structure this problem is multiobjective, i.e. there are two or more competing objectives which have to be minimized at the same time. On the one hand there exists the target that the tumor has to be irradiated sufficiently high such that it is destroyed. On the other hand the surrounding organs and tissue, which is also affected by this treatment, should be spared. Thereby the physician has to weight the risk of the unavoidable damage of the surrounding to the tumor against each other. He can decide on the reduction of the irradiation dose delivered to one organ by allowing a higher dose level in another organ.

Earlier this problem was solved using a weighted sum approach, i.e. the several objectives are multiplied with weights and summarized for getting a single scalar-valued objective. Thereby it is difficult to choose the weights as they have no medical meaning ([114]). Hence the preferred treatment plan can only be found by the physician by a time-consuming trial-and-error-process ([143, p.224],[175]). Therefore, Hamacher and Küfer ([100, 143]) solved the IMRT problem using multiobjective methods. The average number of objectives which have to be considered in these problem formulations vary between 4 and 15. The number of objectives reflects the number of healthy organs – the so-called organs at risk – surrounding the tumor.

Examinations of these problems show that the number of objectives can often be reduced ([142]). This is due to the fact that a treatment plan which leads to a high dose in one organ also leads to a higher dose level in some neighbored organs at risk. Also the sparing of this dominating organ always leads to a sparing of these neighbored organs. Thus only very few organs seem to dominate in their behavior and are truly competing while the others follow in their reaction these dominating organs with a less high impact. Therefore it suffices to consider the dominating organs. This reduces the number of objectives significantly. For instance in the case of the treatment planning for a prostate cancer only two objectives have to be considered.

With respect to this reduced number of competing objectives the aim is to determine a representative approximation of the efficient set. Especially in IMRT the importance of high quality approximations with almost equidistant points is pointed out ([143, p.229]). We apply the algorithms presented in Chap. 4 for reaching this aim. Based on the approximation the physician can choose the preferred irradiation plan. Having the whole solution set available he can not only compare the objective function values of the several solutions, but he can base his decision also on trade-off [173, 174] information. This is the information how a slight improvement of the dose delivered to one organ leads to the rise of the irradiation damage caused in another organ.

6.1 Problem Formulation Using a Bicriteria Approach

We demonstrate the structure of the problems arising in IMRT planning on the special case of a prostate cancer based on the model and data developed and calculated by Küfer et al. ([142]). Here, the tumor is irradiated by five equidistant beams which are a collection of 400 separately controllable beamlets (or pencil beams). We assume that the irradiation geometry is fixed and we concentrate on an optimization of the irradiation intensity. The problem of finding an optimal irradiation geometry is considered for instance in [61, 64, 200] and the references therein.

The relevant part of the patients body is mapped by a computer tomography (CT), see the Figures 6.1 and 6.2. According to the thick-

6.1 Problem Formulation Using a Bicriteria Approach

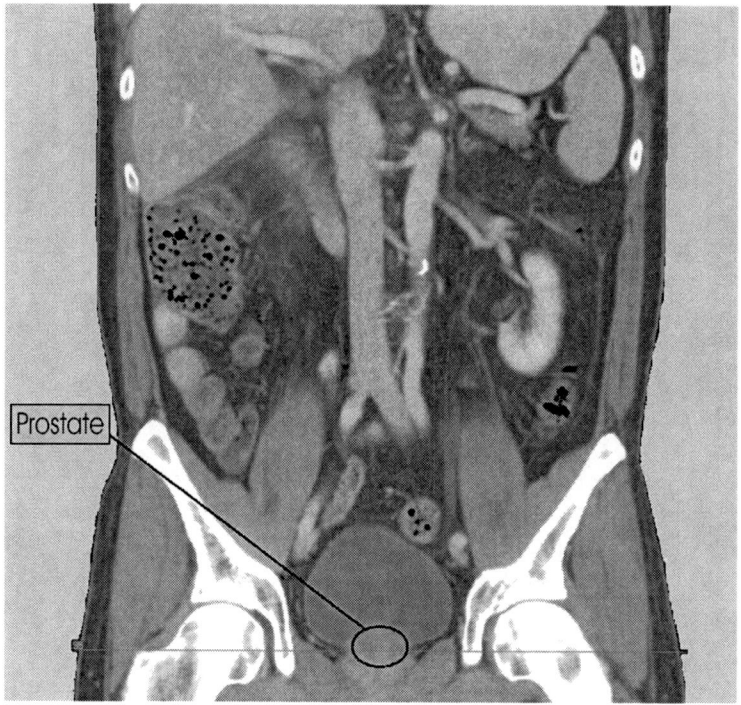

Fig. 6.1. Coronal CT-cut ([130]).

ness of the slices of the CT-cuts the body is dissected in cubes, the so-called voxels v_j, $j = 1, \ldots, 435\,501$.

The large number of voxels ($435\,501$) can be reduced by a clustering method ([143, 198]) to $11\,877$ clusters by collecting those voxels which have the same irradiation stress w.r.t. one irradiation unit. These clusters are denoted as $c_1, \ldots, c_{11877}$. Each of these clusters is allocated to one of the considered volume structures $V_0, \ldots, V_6$ by a physician. In our example these are the tumor (volumes V_0, V_1), the rectum (V_2), the left (V_3) and the right (V_4) hip-bone, the remaining surrounding tissue (V_5) and the bladder (V_6), see Fig. 6.3.

Examinations ([142]) have shown that the bladder (V_6) and the rectum (V_2) are the dominating opponents in their reaction, whereas the other critical organs follow these dominating organs in their stress caused by different irradiation plans. The sparing of the bladder leads to a high dose in the rectum and vice versa.

170   6 Application to IMRT

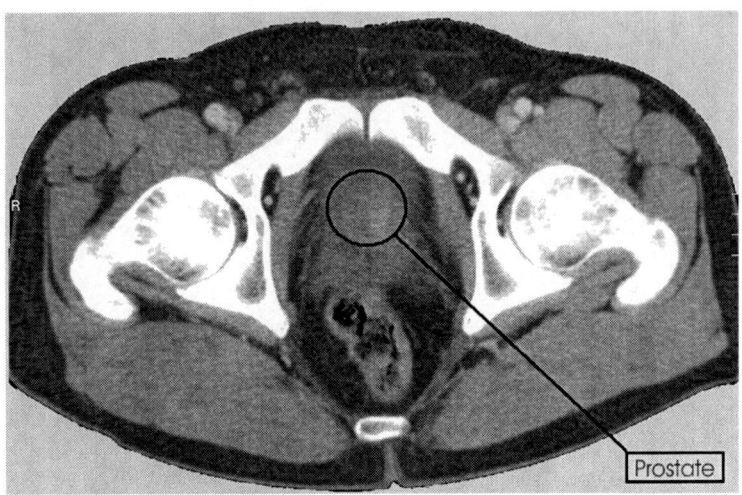

Fig. 6.2. Axial CT-cut ([130]).

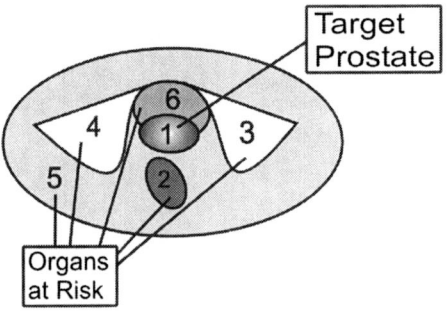

Fig. 6.3. Schematic axial body cut.

The emission of the beamlets B_i ($i \in \{1, \ldots, 400\}$) to the clusters c_j ($j \in \{1, \ldots, 11\,877\}$) at one radiation unit is described by the matrix $P = (P_{ji})_{j=1,\ldots,11\,877, i=1,\ldots,400}$ ([100, p.149], [223, Chap. 3]). Let $x \in \mathbb{R}^{400}$ denote the intensity profile of the beamlets. Then $P_j x$ with P_j the jth row of the matrix P describes the irradiation dose in the cluster c_j caused by the beamlets B_i, $i = 1, \ldots, 400$, for the treatment plan x.

For evaluating and comparing the irradiation stress in the organs we use the concept of the equivalent uniform dose (EUD), see [23]. This uniform dose is defined as that dose level for each organ, which delivered evenly to the entire organ has the same biological effect as the actual irregular irradiation caused by the irradiation intensity x and expressed

6.1 Problem Formulation Using a Bicriteria Approach

by Px has. That value can be calculated based on Nimierko's EUD ([176], see also [37]) using p-norms (here with respect to the clustered voxels) by

$$E_k(x) := \left(\frac{1}{N(V_k)} \sum_{\{j|c_j \in V_k\}} N(c_j) \cdot (P_j x)^{p_k} \right)^{\frac{1}{p_k}},$$

for $k = 2, \ldots, 6$. The deviation of the irregular dose in an organ to the desired limit U_k, which should not be exceeded, is then measured by the convex function

$$\mathrm{EUD}_k(x) := \frac{1}{U_k} E_k(x) - 1,$$

for $k = 2, \ldots, 6$. Here $p_k \in [1, \infty[$ is an organ dependent constant which reflects the physiology of the organ ([223, p.33]). High values are applied for serial organs as the spinal cord. This corresponds to the determination of the maximal dose in a cluster. Small values close to one for p_k are related to parallel structured organs like the liver or the lungs, which corresponds to a mean value. $N(V_k)$ is the number of voxels in organ V_k and $N(c_j)$ is the number of voxels in cluster c_j, thus it is $\sum_{\{j|c_j \in V_k\}} N(c_j) = N(V_k)$. The value U_k as well as p_k are statistical evaluated and can, in our example, be taken from Table 6.1.

Table 6.1. Critical values for the organs at risk.

	number of organ (k)	p_k	U_k	Q_k	$N(V_k)$
rectum	2	3.0	30	36	6 459
left hip-bone	3	2.0	35	42	3 749
right hip-bone	4	2.0	35	42	4 177
remaining tissue	5	1.1	25	35	400 291
bladder	6	3.0	35	42	4 901

A feasible treatment plan has now to fulfill several criteria. First, a dangerous overdosing of the critical tissue should be avoided and thus, the maximal value of Q_k must not be exceeded for all organs at risk V_k, $k = 2, \ldots, 6$, i.e.

$$E_k(x) = U_k(\mathrm{EUD}_k(x) + 1) \leq Q_k, \qquad k = 2, \ldots, 6.$$

These inequalities can be restated as

$$\sum_{\{j|c_j\in V_k\}} N(c_j)(P_jx)^{p_k} \leq Q_k^{p_k} \cdot N(V_k), \qquad k = 2, \ldots, 6.$$

It is also important that the dose in the tumor tissue remains below a maximal value to avoid injuries in the patients body and to achieve homogeneity of the irradiation. Besides, to have the desired effect of destroying the tumor cells a certain curative dose has to be reached in each cluster. Here, we differentiate between the so-called target-tissue V_0 and the boost-tissue V_1, which is tumor tissue that has to be irradiated especially high. These conditions result in the following constraints for every cluster of the target- and the boost-volume:

$$\begin{aligned} l_0(1-\varepsilon_0) \leq P_j x \leq l_0(1+\delta_0), & \quad \forall j \text{ with } c_j \in V_0 \\ \text{and } l_1(1-\varepsilon_1) \leq P_j x \leq l_1(1+\delta_1), & \quad \forall j \text{ with } c_j \in V_1, \end{aligned} \qquad (6.1)$$

where l_0, l_1, ε_0, ε_1, δ_0 and δ_1 are constants given by the physician and tabulated in Table 6.2. Volume V_0 consists of 8 593 clusters while V_1 has 302 clusters. Including nonnegativity constraints for the beamlet intensity this results in the constraint set

$$\begin{aligned} \Omega = \{x \in \mathbb{R}_+^{400} \mid & U_k(\text{EUD}_k(x)+1) \leq Q_k, \quad k = 2, \ldots, 6, \\ & l_0(1-\varepsilon_0) \leq P_j x \leq l_0(1+\delta_0), \quad \forall j \text{ with } c_j \in V_0, \\ & l_1(1-\varepsilon_1) \leq P_j x \leq l_1(1+\delta_1), \quad \forall j \text{ with } c_j \in V_1\} \end{aligned}$$

with 17 795 constraints.

Table 6.2. Critical values for the tumor tissues.

	number of organ (k)	l_k	δ_k	ε_k
target-tissue	0	67	0.11	0.11
boost-tissue	1	72	0.07	0.07

The aims are a minimization of the dose stress in the rectum (V_2) and in the bladder (V_6), as these two healthy organs always have the highest irradiation stress compared to the other organs at risk and a stress reduction for the rectum deteriorates the level for the bladder and vice versa. This leads to the bi-objective optimization problem

$$\min \begin{pmatrix} f_1(x) \\ f_2(x) \end{pmatrix} = \begin{pmatrix} \text{EUD}_6(x) \\ \text{EUD}_2(x) \end{pmatrix}$$

subject to the constraint

$$x \in \Omega$$

6.1 Problem Formulation Using a Bicriteria Approach

with the image space ordered by $K = \mathbb{R}^2_+$.

We apply Alg. 1 with $r = (1,1)^\top$, $\alpha = 0.04$,

$$H := \{y \in \mathbb{R}^2 \mid y_1 = 0\} = \{y \in \mathbb{R}^2 \mid (1,0)y = 0\},$$

i.e. $b = (1,0)^\top$, $\beta = 0$, and we choose $M^1 = 100\,000$ with $M^1 > f_2(x) - f_1(x)$ for all $x \in \Omega$. Then we get, as a result of the Steps 1 and 2, the points

$$f(\bar{x}^1) \approx \begin{pmatrix} 0.0159 \\ 0.2000 \end{pmatrix}$$

with $f_1(\bar{x}^1) = \min_{x \in \Omega} f_1(x)$ and

$$f(\bar{x}^E) \approx \begin{pmatrix} 0.2000 \\ -0.0197 \end{pmatrix}$$

with $f_2(\bar{x}^E) = \min_{x \in \Omega} f_2(x)$. Thus only parameters $a \in H^a$ with

$$H^a = \{y \in \mathbb{R}^2 \mid y_1 = 0,\ y_2 = \lambda \cdot 0.1841 + (1-\lambda) \cdot (-0.2197),\ \lambda \in [0,1]\}$$

have to be considered. With the algorithm the approximation given in Fig. 6.4 with 10 approximation points (connected with lines) is generated.

These points as well as the distances $\delta^i := \|f(x^{i+1}) - f(x^i)\|_2$ between consecutive approximation points are listed in Table 6.3. There, also the EUD-values of the other organs are given.

Finally we can also compare the minimal and maximal equivalent uniform dose value E_k in the organs V_k, $k = 2, \ldots, 6$, over all approximation points with the recommended limits, see Table 6.4.

Instead of using the Pascoletti-Serafini scalarization with the described parameters we can also use the ε-constraint method and Alg. 2. This is of course just a special case of Alg. 1 for special chosen parameters. We solve the scalar optimization problems

$$\min\ \mathrm{EUD}_6(x)$$
$$\text{subject to the constraints}$$
$$\mathrm{EUD}_2(x) \leq \varepsilon,$$
$$x \in \Omega$$

174   6 Application to IMRT

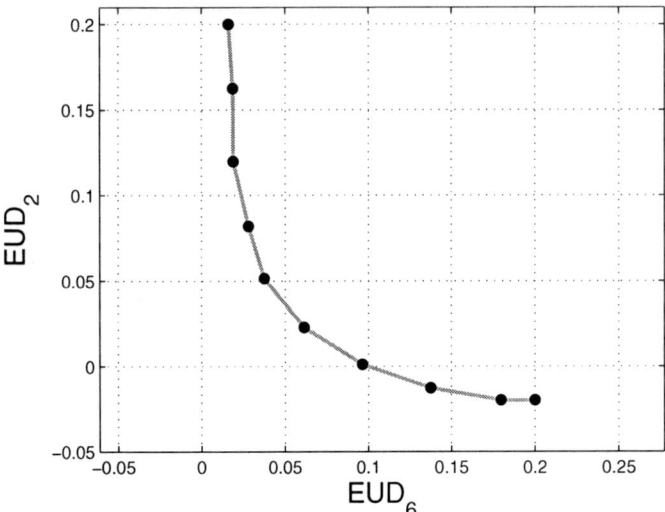

Fig. 6.4. Efficient set and approximation points of the bicriteria optimization problem determined with the adaptive Pascoletti-Serafini method.

Table 6.3. Approximation points of Fig. 6.4 and distances δ^i between them for $\alpha = 0.04$.

approximation point	$i=1$	$i=2$	$i=3$	$i=4$	$i=5$
$\text{EUD}_2(\bar{x}^i)$	0.2000	0.1625	0.1197	0.0819	0.0515
$\text{EUD}_6(\bar{x}^i)$	0.0159	0.0184	0.0187	0.0278	0.0374
δ_i	-	0.0375	0.0429	0.0389	0.0319
$\text{EUD}_3(\bar{x}^i)$	0.1999	0.1998	0.2000	0.2000	0.2000
$\text{EUD}_4(\bar{x}^i)$	0.1984	0.1998	0.2000	0.1999	0.2000
$\text{EUD}_5(\bar{x}^i)$	-0.4250	-0.4292	-0.4334	-0.4342	-0.4314
approximation point	$i=6$	$i=7$	$i=8$	$i=9$	$i=10$
$\text{EUD}_2(\bar{x}^i)$	0.0228	0.0012	-0.0126	-0.0197	-0.0197
$\text{EUD}_6(\bar{x}^i)$	0.0615	0.0964	0.1376	0.1796	0.2000
δ_i	0.0375	0.0411	0.0434	0.0426	0.0204
$\text{EUD}_3(\bar{x}^i)$	0.2000	0.1998	0.1963	0.1776	0.1790
$\text{EUD}_4(\bar{x}^i)$	0.2000	0.1429	0.1125	0.1063	0.1143
$\text{EUD}_5(\bar{x}^i)$	-0.4293	-0.4176	-0.4156	-0.4122	-0.4128

6.1 Problem Formulation Using a Bicriteria Approach

Table 6.4. Extremal values for E_k and recommended limits for the ten approximation points of the approximation of Fig. 6.4.

k	$\min_{i \in \{1,\ldots,10\}} E_k(\bar{x}^i)$	$\max_{i \in \{1,\ldots,10\}} E_k(\bar{x}^i)$	U_k	Q_k
2	29.41	36.00	30	36
3	41.22	42.00	35	42
4	38.72	42.00	35	42
5	14.15	14.69	25	35
6	35.55	42.00	35	42

for parameters $\varepsilon \in \mathbb{R}$. We first solve the problems $\min_{x \in \Omega} f_i(x)$, $i = 1, 2$ and we get, according to Corollary 2.31, that it is sufficient to consider parameters $\varepsilon \in \mathbb{R}$ with $\varepsilon \in [-0.0197, 0.2000]$.

We apply the proposed algorithm with a first-order approximation and a desired distance between the approximation points of $\alpha = 0.04$. This results in the parameters

$$\varepsilon \in \{0.2000, 0.1600, 0.1203, 0.0805, 0.0421, 0.0069, -0.0183, -0.0197\}$$

and the approximation shown in Fig. 6.5. The values of the approximation points and the distances δ^i between consecutive points are tabulated in Table 6.5.

Table 6.5. Values of the approximation points and distances for the approximation of Fig. 6.5.

app.point	$i=1$	$i=2$	$i=3$	$i=4$	$i=5$	$i=6$	$i=7$	$i=8$
$\mathrm{EUD}_2(x^i)$	0.2000	0.1600	0.1203	0.0805	0.0421	0.0069	-0.0183	-0.0197
$\mathrm{EUD}_6(x^i)$	0.0159	0.0164	0.0186	0.0283	0.0425	0.0782	0.1356	0.2000
δ^i	0.0400	0.0398	0.0410	0.0410	0.0501	0.0627	0.0644	-

Based on one of these approximations the physician can choose a treatment plan by weighting the damage to the bladder and the rectum against each other. Besides he can choose an interesting solution and refine around it by using the strategy as given in Step 4 of Alg. 5. Further he can choose a point y determined by interpolation between consecutive approximation points and solve problem (SP(a, r)) to the correspondent parameter, see [223], to get a new approximation point.

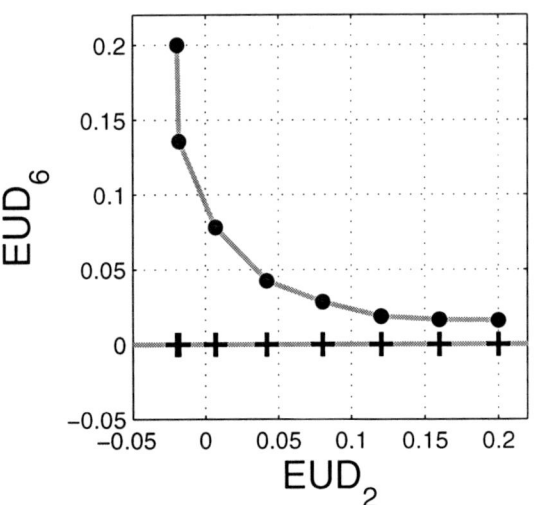

Fig. 6.5. Approximation of the efficient set determined with the adaptive ε-constraint method.

6.2 Problem Formulation Using a Tricriteria Approach

As it turned out that the treatment success depends also on the irradiation homogeneity ([100, p.150]) this aim should be added to the former two objective functions. Thereby the homogeneity of the irradiation is measured by

$$\mathrm{HOM}(x) := \sqrt{\frac{\sum_{\{j|c_j \in V_0\}} N(c_j)(P_j x - l_0)^2 + \sum_{\{j|c_j \in V_1\}} N(c_j)(P_j x - l_1)^2}{N(V_0) + N(V_1)}}$$

with $N(V_0) = 13\,238$ and $N(V_1) = 2\,686$. This results in the multiobjective optimization problem

$$\min \begin{pmatrix} f_1(x) \\ f_2(x) \\ f_3(x) \end{pmatrix} = \begin{pmatrix} \mathrm{EUD}_6(x) \\ \mathrm{EUD}_2(x) \\ \mathrm{HOM}(x) \end{pmatrix}$$

subject to the constraint

$$x \in \Omega$$

with three competing objectives and the ordering cone $K = \mathbb{R}^3_+$.

6.2 Problem Formulation Using a Tricriteria Approach

We solve this problem using Alg. 5. The auxiliary problem is

$$\min \text{HOM}(x)$$
$$\text{subject to the constraints}$$
$$\text{EUD}_6(x) \leq \varepsilon_1,$$
$$\text{EUD}_2(x) \leq \varepsilon_2,$$
$$x \in \Omega.$$

We choose $N^1 = N^2 = 3$. In Step 1 we get $\varepsilon_1^{\min} = 0.0158$, $\varepsilon_1^{\max} = 0.2000$, $\varepsilon_2^{\min} = -0.0141$, and $\varepsilon_2^{\max} = 0.2000$. This results in

$$L_1 = \frac{0.2000 - 0.0158}{3} = 0.0614 \text{ and}$$
$$L_2 = \frac{0.2000 - (-0.0141)}{3} = 0.0714.$$

and thus in the parameter set

$$E := \{\varepsilon \in \mathbb{R}^2 \mid \varepsilon_1 = 0.0158 + (k_1 + \tfrac{1}{2}) L_1, \ k_1 \in \{0, 1, 2\},$$
$$\varepsilon_2 = -0.0141 + (k_2 + \tfrac{1}{2}) L_2, \ k_2 \in \{0, 1, 2\}\}$$
$$= \{\varepsilon \in \mathbb{R}^2 \mid \varepsilon_1 \in \{0.0465, \ 0.1079, \ 0.1693\},$$
$$\varepsilon_2 \in \{0.0216, \ 0.0929, \ 0.1643\}\}.$$

For solving the related scalar optimization problems we use the SQP procedure implemented in Matlab with 600 iterations and a restart after 150 iteration steps. We do not get a solution for the parameter $\varepsilon = (0.0465, 0.0216)$.

We assume a physician chooses certain points and we do a refinement around these points with $k = 1$, i.e. $\bar{n} = 8$, and $\alpha = 0.07$. This results in the refined approximation shown in Fig. 6.6. The determined parameters $(\varepsilon_1, \varepsilon_2) =: (a_1, a_2)$ according to the Steps 2 and 4 are shown in Fig. 6.7. The resulting approximation points are given in Table 6.6.

Due to the high dimension together with the large number of constraints, solving the scalar optimization problems is difficult. Stopping after 600 iterations leads to slightly infeasible solutions. For instance, for a feasible solution x the constraint $\text{EUD}_3(x) \leq 0.2000$ should be satisfied. However $\text{EUD}_3(x^1) = 0.2053$. Further the constraints

$$U_k(\text{EUD}_k(x) + 1) \leq Q_k, \qquad k \in \{2, 3, 4, 5, 6\}$$

are satisfied if

178   6 Application to IMRT

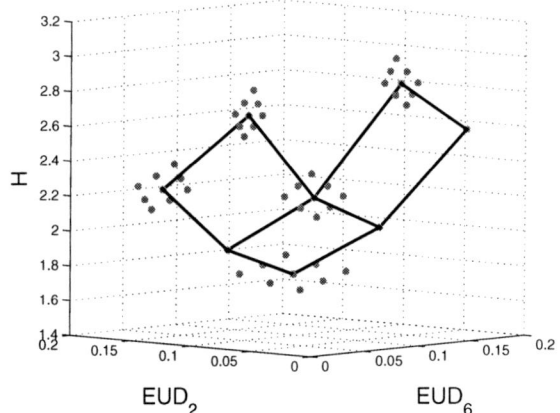

Fig. 6.6. Refined approximation of the IMRT problem. Here $H(\cdot) := \mathrm{HOM}(\cdot)$.

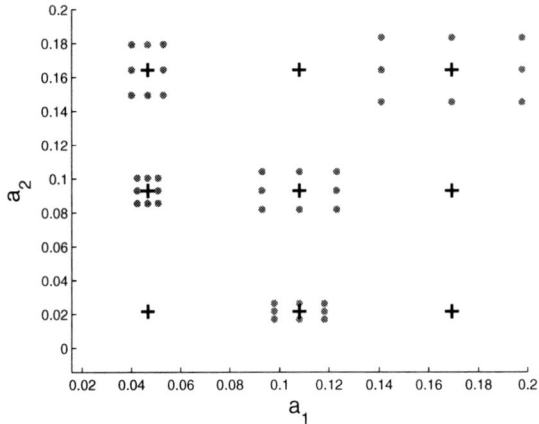

Fig. 6.7. IMRT problem: parameter set of the tricriteria problem formulation.

$$\mathrm{EUD}_k(x) \leq 0.2 \text{ for } k \in \{2,3,4,6\}$$

and

$$\mathrm{EUD}_5(x) \leq 0.4.$$

If these inequalities are fulfilled for a treatment plan, then the irradiation dose in the correspondent organs remains below the maximal value Q_k. If $\mathrm{EUD}_k(x) < 0$, then the irradiation of the organ V_k is

6.2 Problem Formulation Using a Tricriteria Approach

Table 6.6. Values of the approximation points for the tricriteria problem formulation.

App.funkt	$i=1$	$i=2$	$i=3$	$i=4$
$\mathrm{EUD}_2(x^i)$	0.0465	0.0465	0.1079	0.1079
$\mathrm{EUD}_6(x^i)$	0.0929	0.1643	0.0216	0.0929
$\mathrm{HOM}(x^i)$	2.6981	2.2294	2.8866	2.1878
$\mathrm{EUD}_3(x^i)$	0.2053	0.2040	0.2048	0.2008
$\mathrm{EUD}_4(x^i)$	0.2320	0.2035	<0	0.2007
$\mathrm{EUD}_5(x^i)$	<0	<0	<0	<0
App.punkt	$i=5$	$i=6$	$i=7$	$i=8$
$\mathrm{EUD}_2(x^i)$	0.1079	0.1693	0.1693	0.1693
$\mathrm{EUD}_6(x^i)$	0.1643	0.0216	0.0929	0.1643
$\mathrm{HOM}(x^i)$	1.8446	2.5875	1.9813	1.6685
$\mathrm{EUD}_3(x^i)$	0.2005	0.2047	<0	0.2006
$\mathrm{EUD}_4(x^i)$	0.2006	<0	<0	0.2004
$\mathrm{EUD}_5(x^i)$	<0	<0	<0	<0

less then the lower limit U_k, see Table 6.6. The other constraints are $P_j(x) \in [59.63, 74.37]$ for j with $c_j \in V_0$ as well as $P_j x \in [66.96, 77.04]$ for j with $c_j \in V_1$. These constraints are also not always satisfied as

$$\min_{i \in \{1,\ldots,8\}} \min_{\{j|c_j \in V_0\}} P_j x^i = 59.57,$$
$$\max_{i \in \{1,\ldots,8\}} \max_{\{j|c_j \in V_0\}} P_j x^i = 74.43,$$
$$\min_{i \in \{1,\ldots,8\}} \min_{\{j|c_j \in V_1\}} P_j x^i = 66.91,$$
$$\max_{i \in \{1,\ldots,8\}} \max_{\{j|c_j \in V_1\}} P_j x^i = 77.10.$$

This infeasibility can be avoided by using other numerical solvers or by a larger number of iterations.

For an evaluation of the quality of the local approximation refinement we choose for instance the approximation point (0.0465, 0.1643, 0.2294) of the efficient set and calculate the distances between that point and the surrounding refinement points. We get the following 12 distances:

$$0.0697, \ 0.0801, \ 0.0795, \ 0.0814,$$
$$0.0880, \ 0.0679, \ 0.0736, \ 0.0624,$$
$$0.0663, \ 0.0687, \ 0.0640, \ 0.0712$$

with a rounded average value of 0.0727. Recall that we have chosen $\alpha = 0.07$ for the refinement.

Part III

Multiobjective Bilevel Optimization

7
Application to Multiobjective Bilevel Optimization

Based on the results in the previous chapters it is also possible to develop a solution method for nonlinear multiobjective bilevel optimization problems. Bilevel optimization, more generally multilevel optimization, is an active research area in mathematical programming, see the monographs by Dempe [49] and Bard [9] as well as [33, 34, 50] and the bibliography reviews by Calamai and Vicente ([229]) and Dempe ([51]). The applications of bilevel optimization are numerous, for instance in economic development policy, agriculture economics, road network design, oil industry regulation, international water systems and flood control, energy policy, traffic assignment and many more, compare [226] and see also [33, 168].

Many papers have been published in the last two decades about bilevel optimization, but there are only very few of them dealing with multiobjective bilevel problems. Procedures for solving linear multiobjective bilevel problems are presented for instance in [178]. Even less papers are about nonlinear multiobjective bilevel problems: Shi and Xia ([206, 207]) present an interactive method, Osman and Abo-Sinna et al. ([1, 179]) propose the usage of fuzzy set theory for convex problems, and Teng et al. ([222]) give an approach for a convex multiperson multiobjective bilevel problem. The first papers presenting a solution method for non-convex nonlinear multiobjective bilevel optimization problems, which computes approximations of the whole solution set, are [68], and then [129, 195] by Jahn and Schaller.

Bonnel and Morgan examine in [20] a bilevel optimization problem with a vector-valued objective function on the lower level and a scalar-valued objective function on the upper level. They call this problem a semivectorial bilevel optimization problem. In [153] Liou et al. study also such problems and denote them mathematical programs with vector optimization constraints.

Despite multiobjective bilevel optimization has not yet received a broad attention in the literature, these problems are very interesting in the view of applications, see for instance [241]. For getting an idea of this we give an illustrative example. Let us consider a city bus transportation system financed by the public authorities. They have as targets the reduction of the money losses in this non-profitable business as well as the reduction of the number of cars on the streets. The public authorities can decide about the bus ticket price, but this influences the customers in their usage of the buses. The public has maybe several competing objectives, too, as to minimize their transportation time and costs. Hence the usage of the public transportation system can be modeled on the lower level with the bus ticket price as parameter. The solution of the lower level influences then the objective values of the public authorities on the upper level. Such a problem can thus be mapped by multiobjective bilevel optimization.

Multiobjective bilevel optimization problems are also closely related to equilibrium problems and the definition of non-dominated equilibrium solutions, see for instance [8, 30, 35, 171, 177, 205, 240].

7.1 Basic Concepts of Bilevel Optimization

In bilevel optimization, also called two-level optimization, problems are considered where the constraint set of the so-called upper level problem is given by the solution set of a so-called lower level (parametric) optimization problem. The variables of the upper level, the upper level variables $y \in \mathbb{R}^{n_2}$ ($n_2 \in \mathbb{N}$), are the parameters of the lower level problem, and again the solutions $x \in \mathbb{R}^{n_1}$ ($n_1 \in \mathbb{N}$) of the optimization problem on the lower level influence the upper level objective function value. The leader on the higher level controls thus a first set of decision variables, for instance prices or resource allocation, while the follower on the lower level controls a second set of decision variables, for instance production volumes or technology alternatives (compare [226]).

For a constant y let the point $x = x(y)$ be a minimal solution of the optimization problem

$$x(y) \in \Psi(y) := \mathrm{argmin}_x \{f(x,y) \mid (x,y) \in G\} \subset \mathbb{R}^{n_1}$$

parameterized by y with a continuously differentiable function $f \colon \mathbb{R}^{n_1} \times \mathbb{R}^{n_2} \to \mathbb{R}^{m_1}$, $m_1 \in \mathbb{N}$, and $G \subset \mathbb{R}^{n_1} \times \mathbb{R}^{n_2}$. This optimization problem,

7.1 Basic Concepts of Bilevel Optimization

dependent on y, is called lower level problem:

$$\min_{x} f(x, y)$$
$$\text{subject to the constraint} \quad (7.1)$$
$$x \in G(y) := \{x \in \mathbb{R}^{n_1} \mid (x, y) \in G\},$$

The superordinate optimization problem of the upper level is then given by

$$\text{"}\min_{y}\text{"} F(x(y), y)$$
$$\text{subject to the constraints} \quad (7.2)$$
$$x(y) \in \Psi(y),$$
$$y \in \tilde{G}$$

with a continuously differentiable function $F \colon \mathbb{R}^{n_1} \times \mathbb{R}^{n_2} \to \mathbb{R}^{m_2}$, $m_2 \in \mathbb{N}$, and a compact set $\tilde{G} \subset \mathbb{R}^{n_2}$. Here the constraint $y \in \tilde{G}$ is uncoupled from the lower level variable as it is assumed e.g. in [9, p.303], [49, pp.123f.], [51, 52, 85].

A more general formulation of the bilevel problem is reached by allowing the constraint $y \in \tilde{G}$ to depend on the lower level variable x (compare [9, 49, 50, 88, 145, 229]), i.e. to allow $(x, y) \in \tilde{G}$ with $\tilde{G} \subset \mathbb{R}^{n_1} \times \mathbb{R}^{n_2}$ instead of $y \in \tilde{G}$ with $\tilde{G} \subset \mathbb{R}^{n_2}$. We discuss this more general formulation in Sect. 7.6. However, for instance in [26, 153] no constraints on the upper level are considered at all. Notice, that in the first publication which used the term bilevel optimization ([27]) the problems did not involve joint upper level constraints (see also [34]).

We speak of a multiobjective bilevel optimization problem if $m_1 \geq 2$ or $m_2 \geq 2$ and in this book we even assume $m_1, m_2 \geq 2$.

If the minimal solution of the lower level problem (7.1) is not unique, the objective function $F(x(\cdot), \cdot)$ is not well-defined for $y \in \mathbb{R}^{n_2}$. Because of that it is written "min" in (7.2). This difficulty is in some papers avoided by just assuming that the solution of the lower level problem is unique. However in the case of a multiobjective optimization problem ($m_1 \geq 2$) on the lower level this cannot be done any more. In this case of non-uniqueness a common procedure is the so-called optimistic approach ([49], [50, p.5], [52, p.506], [178, p.166]). Then it is assumed that the d.m. of the lower level chooses the minimal solution (for a fixed value of y) which is best for the upper level, thus it is solved

$$\min_{x}\{F(x, y) \mid x \in \Psi(y)\} =: \varphi_0(y).$$

7 Application to Multiobjective Bilevel Optimization

In the case of a vector-valued objective function F the function φ is generally a set-valued map. In the case of a scalar-valued map F ($m_2 = 1$) the bilevel problem using the optimistic approach is then written as

$$\min_y \varphi_0(y)$$

subject to the constraint

$$y \in \tilde{G}.$$

This scalar optimization problem is – w.r.t. the global minimal solutions – equivalent to

$$\min_{x,y} F(x,y)$$

subject to the constraints

$$x \in \Psi(y),\ y \in \tilde{G}.$$

Note that this equivalence is in general only true for global and not for local minimal solutions, see [52, p.507],[53, p.585]. Using the same idea of an optimistic approach for a multiobjective optimization problem we consider in the following the problem

$$\min_{x,y} F(x,y)$$

subject to the constraints

$$x \in \Psi(y),\ y \in \tilde{G}.$$

For the multiobjective optimization problems on the different levels we assume that the partial ordering on the upper level is given by the closed pointed convex cone $K^2 \subset \mathbb{R}^{m_2}$ and on the lower level by the closed pointed convex cone $K^1 \subset \mathbb{R}^{m_1}$. Further we assume that for any $y \in \tilde{G}$ of the upper level there exists a minimal solution of the lower level problem.

7.2 Induced Set Approximation

Using the optimistic approach we hence consider the multiobjective bilevel optimization problem

$$\min_{x,y} F(x,y)$$

subject to the constraints

$$x \in \mathcal{M}_y(f(G), K^1),$$
$$y \in \tilde{G}$$

(7.3)

7.2 Induced Set Approximation

with $\mathcal{M}_y(f(G), K^1) := \mathcal{M}(f(G(y), y), K^1)$ the set of K^1-minimal points of the multiobjective optimization problem (7.1) parameterized by y. The constraint set Ω of the upper level problem, also called induced set, is then given by

$$\Omega := \{(x, y) \in \mathbb{R}^{n_1} \times \mathbb{R}^{n_2} \mid x \in \mathcal{M}_y(f(G), K^1),\ y \in \tilde{G}\}.$$

We show that the induced set Ω is equivalent to the set of $\hat{K}$-minimal points of the multiobjective optimization problem

$$\min_{x,y} \hat{f}(x,y) := \begin{pmatrix} f(x,y) \\ y \end{pmatrix}$$
$$\text{subject to the constraints} \qquad\qquad (7.4)$$
$$(x, y) \in G,$$
$$y \in \tilde{G}$$

with $\hat{K} := K^1 \times \{0_{n_2}\} \subset \mathbb{R}^{m_1} \times \mathbb{R}^{n_2}$.

Theorem 7.1. *Let $\hat{\mathcal{M}}$ be the set of $\hat{K}$-minimal points of the multiobjective optimization problem (7.4) with $\hat{K} = K^1 \times \{0_{n_2}\}$. Then $\Omega = \hat{\mathcal{M}}$.*

Proof. We have the equivalences

$(\bar{x}, \bar{y}) \in \Omega \Leftrightarrow \bar{x} \in \mathcal{M}_{\bar{y}}(f(G), K^1) \quad \wedge \quad \bar{y} \in \tilde{G}$
$\Leftrightarrow (\nexists\ x \in G(\bar{y}) \text{ with } f(\bar{x}, \bar{y}) \in f(x, \bar{y}) + K^1 \setminus \{0_{m_1}\})$
$\qquad \wedge\ \bar{y} \in \tilde{G}\ \wedge\ \bar{x} \in G(\bar{y})$
$\Leftrightarrow (\nexists\ (x, y) \in G \text{ with } f(\bar{x}, \bar{y}) \in f(x, y) + K^1 \setminus \{0_{m_1}\}$
$\qquad \wedge\ y = \bar{y})\ \wedge\ \bar{y} \in \tilde{G}\ \wedge\ (\bar{x}, \bar{y}) \in G$
$\Leftrightarrow (\nexists\ (x, y) \in G \text{ with}$
$\qquad \begin{pmatrix} f(\bar{x}, \bar{y}) \\ \bar{y} \end{pmatrix} \in \begin{pmatrix} f(x, y) \\ y \end{pmatrix} + (K^1 \times \{0_{n_2}\}) \setminus \{0_{m_1+n_2}\})$
$\qquad \wedge\ \bar{y} \in \tilde{G}\ \wedge\ (\bar{x}, \bar{y}) \in G$
$\Leftrightarrow (\nexists (x, y) \in G \text{ with } \hat{f}(\bar{x}, \bar{y}) \in \hat{f}(x, y) + \hat{K} \setminus \{0_{m_1+n_2}\})$
$\qquad \wedge\ \bar{y} \in \tilde{G}\ \wedge\ (\bar{x}, \bar{y}) \in G$
$\Leftrightarrow (\bar{x}, \bar{y}) \in \hat{\mathcal{M}}.$

□

Hence, if we are able to determine the solution set of the multiobjective optimization problem (7.4) we have already the constraint set

7 Application to Multiobjective Bilevel Optimization

of the upper level problem, which we can solve then. The upper level problem is then reduced to $\min_{x,y}\{F(x,y) \mid (x,y) \in \hat{\mathcal{M}}\}$.

We cannot determine the whole solution set of the problem (7.4), but we can calculate an approximation of this set. Based on sensitivity information, we can refine this approximation dependent on the behavior of the upper level function.

For determining single solution points of the problem (7.4) we use the scalarization according to Pascoletti and Serafini and we get the scalar optimization problem (SP($\hat{a}, \hat{r}$))

$$\min_{t,x,y} t$$

subject to the constraints

$$\hat{a} + t\hat{r} - \hat{f}(x,y) \in \hat{K}, \qquad (7.5)$$
$$(x,y) \in G,$$
$$y \in \tilde{G},$$
$$t \in \mathbb{R}$$

with $\hat{a} \in \mathbb{R}^{m_1+n_2}$, $\hat{r} \in \hat{K} = K^1 \times \{0_{n_2}\}$. In the following we assume, for getting an easier notation, $e_{m_1} \in K^1$ with e_{m_1} the m_1th unit vector in $\mathbb{R}^{m_1}$. This is for instance satisfied for $K^1 = \mathbb{R}_+^{m_1}$, i.e. for the natural ordering. Then it is sufficient (see Theorem 2.11) to consider only parameters

$$\hat{a} := \begin{pmatrix} a \\ \tilde{a} \end{pmatrix} \in \hat{H} = \{x \in \mathbb{R}^{m_1+n_2} \mid x_{m_1} = 0\}, \ \hat{r} = \begin{pmatrix} r \\ 0_{n_2} \end{pmatrix} \text{ with } r = e_{m_1} \qquad (7.6)$$

with $a \in \mathbb{R}^{m_1}$ and $\tilde{a} \in \mathbb{R}^{n_2}$. In the next theorem we show that with these parameters a point $(\bar{t}, \bar{x}, \bar{y})$ is a minimal solution of (SP($\hat{a}, \hat{r}$)), if $(\bar{t}, \bar{x})$ is a minimal solution of the problem (SP($a, r, \tilde{a}$)) defined by

$$\min_{t,x} t$$

subject to the constraints

$$a + tr - f(x, \tilde{a}) \in K^1, \qquad (7.7)$$
$$(x, \tilde{a}) \in G,$$
$$t \in \mathbb{R}$$

with $a \in H := \{x \in \mathbb{R}^{m_1} \mid x_{m_1} = 0\}$, $\tilde{a} \in \tilde{G}$, and $r = e_{m_1}$. Problem (SP($\hat{a}, \hat{r}$)) has, for parameters as in (7.6), no minimal solution for $\tilde{a} \notin$

7.2 Induced Set Approximation

$\tilde{G}$. Thus it is only interesting to consider the problem $(SP(\hat{a}, \hat{r}))$ for $\hat{a} = (a, \tilde{a})$ with $\tilde{a} \in \tilde{G}$ and then we can ignore the constraint $y \in \tilde{G}$.

Theorem 7.2. *We consider the optimization problems $(SP(\hat{a}, \hat{r}))$ and $(SP(a, r, \tilde{a}))$ with*

$$G = \{(x, y) \in \mathbb{R}^{n_1+n_2} \mid g(x, y) \in C\}$$

with a continuously differentiable function $g \colon \mathbb{R}^{n_1+n_2} \to \mathbb{R}^p$, $p \in \mathbb{N}$, and a convex cone $C \subset \mathbb{R}^p$. Let the point $(\bar{t}, \bar{x})$ be a minimal solution of $(SP(a, r, \tilde{a}))$ with $a \in H$, $r = e_{m_1}$, and $\tilde{a} \in \tilde{G}$ with Lagrange multipliers $\mu \in (K^1)^*$ (with $(K^1)^*$ the dual cone to K^1) to the constraint $a + tr - f(x, \tilde{a}) \in K^1$ and $\nu \in C^*$ to the constraint $g(x, y) \in C$.

Then $(\bar{t}, \bar{x}, \tilde{a})$ is a minimal solution of $(SP(\hat{a}, \hat{r}))$ with $\hat{a}$ and $\hat{r}$ as in (7.6) and Lagrange multipliers $(\mu, \tilde{\mu}) \in \hat{K}^* = (K^1)^* \times \mathbb{R}^{n_2}$ with

$$\tilde{\mu} := -\sum_{i=1}^{m_1} \mu_i \nabla_y f_i(\bar{x}, \tilde{a}) + \sum_{j=1}^{p} \nu_j \nabla_y g_j(\bar{x}, \tilde{a}) \tag{7.8}$$

to the constraint $\hat{a} + t\hat{r} - \hat{f}(x, y) \in \hat{K}$ and Lagrange multiplier $\nu \in C^*$ to the constraint $g(x, y) \in C$.

Proof. Let $(\bar{t}, \bar{x})$ be a minimal solution of $(SP(a, r, \tilde{a}))$. Then we have for the related Lagrange function

$$\mathcal{L}(\bar{t}, \bar{x}, \mu, \nu) := \bar{t} - \mu^\top (a + \bar{t}r - f(\bar{x}, \tilde{a})) - \nu^\top g(\bar{x}, \tilde{a})$$

that

$$\nabla_{(t,x)} \mathcal{L}(\bar{t}, \bar{x}, \mu, \nu) = \begin{pmatrix} 1 \\ 0_{n_1} \end{pmatrix} + \sum_{i=1}^{m_1-1} \mu_i \begin{pmatrix} 0 \\ \nabla_x f_i(\bar{x}, \tilde{a}) \end{pmatrix}$$
$$+ \mu_{m_1} \begin{pmatrix} -1 \\ \nabla_x f_{m_1}(\bar{x}, \tilde{a}) \end{pmatrix} - \sum_{j=1}^{p} \nu_j \begin{pmatrix} 0 \\ \nabla_x g_j(\bar{x}, \tilde{a}) \end{pmatrix}$$
$$= 0_{n_1+1}.$$

$$\tag{7.9}$$

We consider now the constraint $\hat{a} + t\hat{r} - \hat{f}(x, y) \in \hat{K}$ of the problem $(SP(\hat{a}, \hat{r}))$ with $\hat{K} = K^1 \times \{0_{n_2}\}$. This constraint is equivalent to the constraints

$$a + tr - f(x, y) \in K^1$$

and

$$\tilde{a} - y \in \{0_{n_2}\}.$$

7 Application to Multiobjective Bilevel Optimization

Hence it follows $\tilde{a} = y$ and we conclude for $\tilde{a} \in \tilde{G}$ immediately, that $(\bar{t}, \bar{x}, \bar{y})$ with $\bar{y} = \tilde{a}$ is a minimal solution of $(\mathrm{SP}(\hat{a}, \hat{r}))$. The Lagrange function $\hat{\mathcal{L}}$ to problem $(\mathrm{SP}(\hat{a}, \hat{r}))$ is defined by (with $\hat{\nu} \in \hat{K}^*$, $\hat{\mu} \in C^*$)

$$\hat{\mathcal{L}}(\bar{t}, \bar{x}, \bar{y}, \hat{\mu}, \hat{\nu}) := \bar{t} - \hat{\mu}^\top (\hat{a} + \bar{t}\hat{r} - \hat{f}(\bar{x}, \bar{y})) - \hat{\nu}^\top g(\bar{x}, \bar{y})$$

and thus together with $\bar{y} = \tilde{a}$ and $\hat{f}_{m_1+i}(x, y) = y_i$, $i = 1, \ldots, n_2$, we get

$$\nabla_{(t,x,y)} \hat{\mathcal{L}}(\bar{t}, \bar{x}, \bar{y}, \hat{\mu}, \hat{\nu}) = \begin{pmatrix} 1 \\ 0_{n_1} \\ 0_{n_2} \end{pmatrix} + \sum_{i=1}^{m_1-1} \hat{\mu}_i \begin{pmatrix} 0 \\ \nabla_x f_i(\bar{x}, \tilde{a}) \\ \nabla_y f_i(\bar{x}, \tilde{a}) \end{pmatrix}$$

$$+ \hat{\mu}_{m_1} \begin{pmatrix} -1 \\ \nabla_x f_{m_1}(\bar{x}, \tilde{a}) \\ \nabla_y f_{m_1}(\bar{x}, \tilde{a}) \end{pmatrix} + \sum_{i=1}^{n_2} \hat{\mu}_{m_1+i} \begin{pmatrix} 0 \\ 0_{n_1} \\ e_i \end{pmatrix}$$

$$- \sum_{j=1}^{p} \hat{\nu}_j \begin{pmatrix} 0 \\ \nabla_x g_j(\bar{x}, \tilde{a}) \\ \nabla_y g_j(\bar{x}, \tilde{a}) \end{pmatrix}.$$

Here e_i denotes the ith unit vector in $\mathbb{R}^{n_2}$. With (7.9) we get for $\hat{\mu} := (\mu, \tilde{\mu})$ and $\hat{\nu} := \nu$

$$\nabla_{(t,x)} \hat{\mathcal{L}}(\bar{t}, \bar{x}, \bar{y}, (\mu, \tilde{\mu}), \nu) = 0_{n_1+1}$$

and by setting $\tilde{\mu}$ as in (7.8) we conclude

$$\nabla_y \hat{\mathcal{L}}(\bar{t}, \bar{x}, (\mu, \tilde{\mu}), \nu) = \sum_{i=1}^{m_1} \mu_i \nabla_y f_i(\bar{x}, \tilde{a}) + \sum_{i=1}^{n_2} \tilde{\mu}_i e_i - \sum_{j=1}^{p} \nu_j \nabla_y g_j(\bar{x}, \tilde{a}) = 0_{n_2}.$$

Further, as μ and ν are Lagrange multipliers to the problem $(\mathrm{SP}(a, r, \tilde{a}))$, it follows

$$\mu^\top (a + \bar{t}r - f(\bar{x}, \tilde{a})) = 0 \text{ and } \nu^\top g(\bar{x}, \tilde{a}) = 0.$$

For the problem $(\mathrm{SP}(\hat{a}, \hat{r}))$ it is for $\hat{\mu} = (\mu, \tilde{\mu})$ because of $\tilde{a} = y$

$$\hat{\mu}^\top (\hat{a} + \bar{t}\hat{r} - \hat{f}(\bar{x}, \bar{y})) = \mu^\top (a + \bar{t}r - f(\bar{x}, y)) + \tilde{\mu}^\top (\tilde{a} - y) = 0$$

and thus $\hat{\mu} = (\mu, \tilde{\mu})$ and ν are Lagrange multipliers to the point $(\bar{t}, \bar{x})$ for the problem $(\mathrm{SP}(\hat{a}, \hat{r}))$, too. □

7.2 Induced Set Approximation

Of course equality constraints for the set G can be included also. Searching for solutions of the problem (SP($\hat{a},\hat{r}$)) with $\hat{a}=(a,\tilde{a})\in\hat{H}$ and $\hat{r}=(r,0_{n_2})$ can thus be replaced by solving (SP(a,r,y)) for $a\in H$ and $y=\tilde{a}\in\tilde{G}$.

Problem (SP(a,r,y)) is just the Pascoletti-Serafini scalarization applied to problem (7.1) for $y=\tilde{a}$. The connection described in the previous theorem will be important for the application of the sensitivity results.

For determining an approximation of the solution set of (7.4) we proceed as follows: We solve the problem (SP($\hat{a},\hat{r}$)) for a choice of parameters $\hat{a},\hat{r}$ as in (7.6) with $\tilde{a}\in\tilde{G}$. However instead of solving the problem (SP($\hat{a},\hat{r}$)) directly we switch to the problem (SP($a,r,\tilde{a}$)) with parameters $a\in H$, $r=e_{m_1}$, and $\tilde{a}\in\tilde{G}$. The aim is to cover the whole solution set of the problem (7.4). For achieving this we discretize the set $\tilde{G}$ with equal distances, e.g. for $\tilde{G}=[c,d]\subset\mathbb{R}$ by $y^1:=c\leq y^2:=y^1+\beta\leq y^3:=y^1+2\beta\leq\ldots\leq y^{n^y}:=y^1+(n^y-1)\beta\leq d$ ($\beta\in\mathbb{R}_+$, $n^y\in\mathbb{N}$) and solve for any $\tilde{a}=y^k$ of this discretization the problem (SP($a,r,\tilde{a}$)) for a variation of the remaining parameters a and r. This is equivalent to determine the solution set of the lower level problem (7.1) for the parameter $y=y^k$. As already mentioned we can only determine an approximation of this solution set. The aim is an approximation with a high quality, i.e. with equidistant points in the parameter space (and not in the image space).

For the remaining parameters a and r we have seen in Theorem 2.11 that it is sufficient to choose the parameter r constant and to choose the parameter a from a hyperplane. Instead of choosing the parameter a for instance equidistantly from the hyperplane we get better results using sensitivity information for an adaptive controlled choice. For the case $K^1=\mathbb{R}_+^{m_1}$ and $G=\{(x,y)\in\mathbb{R}^{n_1+n_2}\mid g_j(x,y)\geq 0,\ j=1,\ldots,p\}$ (i.e. $C=\mathbb{R}_+^p$) we get such sensitivity results by applying Theorem 3.16 in the non-degenerated case and Theorem 3.17 in the degenerated case.

For our numerical method we assume that all necessary assumptions for the sensitivity results are at least locally satisfied. Let $y=y^k$ be constant. As we used to do it in Chap. 4 we determine an approximation by choosing several parameters $a\in H$, e.g. with $a_1^0\leq a_1^0\leq a_1^2\leq\ldots$, and by solving the problem (SP(a,r,y^k)) for these parameters. Our aim is to have for the minimal solutions (t^i,x^i) of the problems (SP(a^i,r,y^k)) for $i=1,2,\ldots$, for a given distance $\alpha>0$:

$$\|x^{i+1}-x^i\|\approx\alpha. \qquad (7.10)$$

7 Application to Multiobjective Bilevel Optimization

We restrict ourselves in the following to the bicriteria case on the lower level, i.e. $m_1 = 2$, and further $K = \mathbb{R}_+^2$. Then, applying the results of Theorem 3.16 (assuming non-degeneracy, otherwise Theorem 3.17), we get for

$$a := \begin{pmatrix} a_1^0 \\ 0 \end{pmatrix} + \lambda \cdot \begin{pmatrix} 1 \\ 0 \end{pmatrix} \in H$$

($\lambda > 0$) and the constant parameter $r = r^0$

$$\phi(a, r) \approx \phi(a^0, r) + M^{-1} N \begin{pmatrix} a - a^0 \\ 0_2 \end{pmatrix}$$

and thus

$$\begin{pmatrix} t(a,r) \\ x(a,r) \\ \mu(a,r) \\ \nu(a,r) \end{pmatrix} \approx \begin{pmatrix} t^0 \\ x^0 \\ \mu^0 \\ \nu^0 \end{pmatrix} + \lambda M^{-1} N \begin{pmatrix} 1 \\ 0 \\ 0 \\ 0 \end{pmatrix}$$

for (t^0, x^0) a minimal solution of the reference problem $(\mathrm{SP}(a^0, r, y^k))$ with Lagrange multipliers μ^0, ν^0. For $\tilde{v} = (1, 0, 0, 0)^\top$ let $(M^{-1} N \tilde{v})\big|_x$ denote the vector consisting of the second to the $(n_1 + 1)$th entry of the vector $M^{-1} N \cdot \tilde{v}$ (correspondent to the entries related to x). Then we get

$$\begin{aligned}
\|x - x^0\| &= \|x(a, r) - x(a^0, r)\| \\
&= \|x(a^0 + \lambda (1,0)^\top, r) - x(a^0, r)\| \\
&\approx \|x^0 + \lambda (M^{-1} N \tilde{v})\big|_x - x^0\| \\
&= |\lambda| \, \|(M^{-1} N \tilde{v})\big|_x\|.
\end{aligned}$$

Thus we have (7.10) e.g. for $i = 0$ approximately fulfilled for $a^1 := a^0 + \bar{\lambda} \cdot (1,0)^\top$ with

$$\bar{\lambda} := \frac{\alpha}{\|(M^{-1} N \tilde{v})\big|_x\|}. \tag{7.11}$$

Having determined the parameter a^1 we can solve the problem $(\mathrm{SP}(a^1, r, y^k))$ and repeat this procedure for calculating a^2 and so on till an approximation of the solution set is generated by this adaptive parameter control. Notice that the aim are equidistant points in the parameter space and not in the image space.

This procedure has to be done for all discretization points y^k of $\tilde{G}$ and then, for $(\bar{t}, \bar{x})$ a minimal solution of $(\mathrm{SP}(a, r, \tilde{a}))$ with $\tilde{a} = y^k$, the point $(\bar{x}, y^k)$ is an approximation point of the set $\hat{\mathcal{M}}$ and hence of the induced set Ω. Let $A^{0,k}$ denote the set of points $(\bar{x}, y^k)$ gained by solving the problem (SP($a, r, \tilde{a}$)) for $\tilde{a} = y^k$ and for several parameters a. The set

$$A^0 := \bigcup_{k=1}^{n^y} A^{0,k}$$

is the searched approximation of the induced set Ω.

Note that by using the scalarizations (SP($\hat{a}, \hat{r}$)) and (SP($a, r, \tilde{a}$)) respectively we do not always get $\hat{K}$- and K^1-minimal solutions respectively of the related multiobjective optimization problem. We have only that any minimal solution of (SP($a, r, \tilde{a}$)) is weakly K^1-minimal. As for (SP($\hat{a}, \hat{r}$)) we have int($\hat{K}$) = int($K^1 \times \{0_{n_2}\}$) = $\emptyset$, we cannot apply the notion of weak minimality here. Instead, for $\hat{r} \in L(\hat{K})$, a minimal solution of (SP($\hat{a}, \hat{r}$)) is an at least minimal point w.r.t. the intrinsic core icr($\hat{K}$) $\cup \{0_{m_1+n_2}\}$ = (icr(K^1) $\cup \{0_{n_2}\}$) $\times \{0_{n_2}\}$ (see p.11). Of course the set of $\hat{K}$-minimal points is included in the set of minimal points w.r.t. icr($\hat{K}$) $\cup \{0_{m_1+n_2}\}$.

7.3 Induced Set Refinement

Now we can evaluate the points approximating the induced set Ω with the upper level objective function F. Hence, for all points $(x, y) \in A^0$ we calculate the points $F(x, y)$. Just for better visualization possibilities we assume in the following $m_2 = 2$ and $K^2 = \mathbb{R}^2_+$, too. We are only interested in the non-dominated points of the set $\{F(x, y) \mid (x, y) \in A^0\}$, i.e. in the points $(\bar{x}, \bar{y}) \in A^0$, so that there exists no point $(x', y') \in A^0$ with $F(x', y') \neq F(\bar{x}, \bar{y})$ and with

$$F_i(x', y') \leq F_i(\bar{x}, \bar{y}), \quad i = 1, 2.$$

We denote the set of non-dominated points as $\mathcal{M}(F(A^0), \mathbb{R}^2_+)$.

For a not too strict selection of points it can be a better concept to select only the ε-EP-minimal points of the set A^0, because the ε-efficient points are also close to the efficient set, in which we are interested, and hence deliver useful information, too (compare Definition 1.15 and [125, p.3]). The determination of all non-dominated points can be very expensive, if the set A^0 consists of many points. Then so-called Pareto filters ([160, p.1193], [161, p.730]) deliver implementable algorithms. The costs can be reduced by using the method of Graef and Younes, see

[112, p.14], [124, p.337], [242], or, extended with a backward iteration, the Jahn-Graef-Younes method, described in [125, pp.4f]. With the set $\mathcal{M}(F(A^0), \mathbb{R}_+^2)$ we have a first approximation of the solution set of the multiobjective bilevel problem.

For improving this approximation of the solution set we refine the discretization of the constraint set in a neighborhood of the already found solutions. We can do this for all points of the set $\mathcal{M}(F(A^0), \mathbb{R}_+^2)$, or, if the d. m. of the upper level problem is interested in some special chosen efficient points, only in their neighborhood. A third possibility is to consider only those points for which the approximation of the efficient set of the upper level problem is not accurate enough, i. e. only those points where the distance to the next neighbor point, mapped with the function F, is not small enough.

For obtaining a refinement around the special chosen points we use again sensitivity information for determining the parameter a of the lower level scalarization in such a way that the refinement points have nearly a predefined equal distance. For any point $(x, y) \in \mathcal{M}(F(A^0), \mathbb{R}_+^2)$ there is some $t \in \mathbb{R}$ so that the point (t, x) is a minimal solution of $(\text{SP}(a, r, \tilde{a}))$ for $\tilde{a} = y$. It is now important that according to Theorem 7.2 the point (t, x, y) is then a minimal solution of $(\text{SP}(\hat{a}, \hat{r}))$, too, and the Lagrange multipliers are also known, if the Lagrange multipliers of the problem $(\text{SP}(a, r, \tilde{a}))$ are given. Problem $(\text{SP}(\hat{a}, \hat{r}))$ is the scalarization of the problem (7.4). Hence we are interested in the dependence of the minimal solutions of the problem $(\text{SP}(\hat{a}, \hat{r}))$ on the parameter $\hat{a}$. Thereby we include the variable y in our sensitivity and thus in our distance considerations.

For instance for $K^1 = \mathbb{R}_+^2$, $n_2 = 1$ (i.e. $y \in \mathbb{R}$), $\hat{r} = (0, 1, 0)^\top \in K^1 \times \{0\}$ and $\hat{a} = (a_1, a_2, a_3)^\top$ we can rewrite the problem $(\text{SP}(\hat{a}, \hat{r}))$ with the following equality and inequality constraints:

$$\min_{t,x,y} t$$

subject to the constraints

$$\begin{aligned} a_1 &- f_1(x, y) \geq 0, \\ a_2 + t &- f_2(x, y) \geq 0, \\ a_3 &\quad -y = 0, \\ &(x, y) \in G, \\ &y \in \tilde{G} = [c, d], \\ &t \in \mathbb{R}. \end{aligned} \tag{7.12}$$

Let $G := \{(x,y) \in \mathbb{R}^{n_1+1} \mid g_j(x,y) \geq 0, \ j = 1,\ldots,p\}$. For simplicity we discuss in the following only this special case. Applying Corollary 3.2.1 from [79] we get, similar to Theorem 3.16, the following sensitivity results for the minimal solutions of the problem (7.12).

Theorem 7.3. *We consider the problem (7.12) with twice continuously differentiable functions $f_1, f_2, g_j, j = 1,\ldots,p$. Let (t^0, x^0, y^0) be a local minimal solution of the reference problem (7.12) w. r. t. the parameter $\hat{a}^0 = (a_1^0, a_2^0, a_3^0)^\top$ with Lagrange multipliers $\mu_1^0, \mu_2^0 \in \mathbb{R}_+$, $\mu_3^0 \in \mathbb{R}, \nu^0 \in \mathbb{R}_+^p$. Assume that the constraints are non-degenerated and that the gradients w. r. t. (t,x,y) of the active constraints are linearly independent. Assume there is a scalar $\alpha > 0$ so that for the Hessian of the Lagrange function $\mathcal{L}$ in the point (t^0, x^0, y^0) it holds*

$$(t, x^\top, y) \nabla^2_{(t,x,y)} \mathcal{L}(t^0, x^0, y^0, \mu^0, \nu^0) \begin{pmatrix} t \\ x \\ y \end{pmatrix} \geq \alpha \left\| \begin{pmatrix} t \\ x \\ y \end{pmatrix} \right\|^2$$

for all

$(t, x, y) \in \{(t, x, y) \in \mathbb{R}^{n_1+2} \mid \hat{r}_i t = \nabla_x f_i(x^0, y^0)^\top x \text{ if } \mu_i^0 > 0 \text{ for } i \in \{1,2\},$
$\nabla_x g_j(x^0, y^0)^\top x = 0 \text{ if } \nu_j^0 > 0 \text{ for } j \in \{1,\ldots,p\}, \ y = 0\}.$

Then (t^0, x^0, y^0) is a local unique minimal solution of (7.12) for the parameter $\hat{a}^0$ and there is a $\delta > 0$ so that the function $\phi \colon N(\hat{a}^0) \to B_\delta(t^0, x^0, y^0) \times B_\delta(\mu^0, \nu^0)$ (for $N(\hat{a}^0)$ a neighborhood of $\hat{a}^0$),

$$\phi(\hat{a}) := (t(\hat{a}), x(\hat{a}), y(\hat{a}), \mu(\hat{a}), \nu(\hat{a})),$$

has the following first order approximation

$$\phi(\hat{a}) = \phi(\hat{a}^0) + \hat{M}^{-1}\hat{N}\left(\hat{a} - \hat{a}^0\right) + o\left(\|\hat{a} - \hat{a}^0\|\right)$$

in the point (t^0, x^0, y^0) with $\hat{M} :=$

$$\begin{pmatrix} \nabla^2_{(t,x,y)} \mathcal{L} & 0 & -1 & 0 & 0 & & 0 \\ & \nabla_{(x,y)} f_1 & \nabla_{(x,y)} f_2 & e_{(n_1+1)} & -\nabla_{(x,y)} g_1 & \cdots & -\nabla_{(x,y)} g_p \\ \mu_1^0(0, -\nabla_{(x,y)} f_1) & k_1 & 0 & 0 & 0 & \cdots & 0 \\ \mu_2^0(1, -\nabla_{(x,y)} f_2) & 0 & k_2 & 0 & 0 & \cdots & 0 \\ \mu_3^0(0, -e_{(n_1+1)}^\top) & 0 & 0 & 0 & 0 & \cdots & 0 \\ \nu_1^0(0, \nabla_{(x,y)} g_1) & 0 & 0 & 0 & g_1 & \cdots & 0 \\ \vdots & \vdots & \vdots & \vdots & \vdots & \ddots & \vdots \\ \nu_p^0(0, \nabla_{(x,y)} g_p) & 0 & 0 & 0 & 0 & \cdots & g_p \end{pmatrix}$$

196   7 Application to Multiobjective Bilevel Optimization

with $k_1 := \hat{a}_1 - f_1(x^0, y^0)$, $k_2 := \hat{a}_2 + t^0 - f_2(x^0, y^0)$, $g_i := g_i(x^0, y^0)$, $i = 1, \ldots, p$, and

$$\hat{N} := \left[0_{3\times(n_1+2)}, -\mu_1^0 \begin{pmatrix} 1 \\ 0 \\ 0 \end{pmatrix}, -\mu_2^0 \begin{pmatrix} 0 \\ 1 \\ 0 \end{pmatrix}, -\mu_3^0 \begin{pmatrix} 0 \\ 0 \\ 1 \end{pmatrix}, 0_{3\times p} \right]^\top.$$

For the degenerate case we get a correspondent result by adapting Theorem 3.17.

Let us assume that we want to refine in the neighborhood of the point $(x^0, y^0) \in \mathcal{M}(F(A^0), \mathbb{R}_+^2)$. Then, because of $(x^0, y^0) \in A^0$, there is some $t^0 \in \mathbb{R}$ and a parameter $\hat{a}^0 = (a_1^0, a_2^0, a_3^0)^\top \in \hat{H} \subset \mathbb{R}^3$ so that (t^0, x^0, y^0) is a minimal solution of (7.12). Assume $\hat{\mu}^0 = (\mu_1^0, \mu_2^0, \mu_3^0) \in \mathbb{R}^3$, $\nu^0 \in \mathbb{R}^p$ are the related Lagrange multipliers using the results of Theorem 7.2. We want to find now new parameters $\hat{a} \in \hat{H}$, for instance by

$$\hat{a} := \hat{a}^0 + s^1 \cdot v^1 + s^2 \cdot v^2, \quad s^1, s^2 \in \mathbb{R}, \; v^1 := (1, 0, 0)^\top, \; v^2 := (0, 0, 1)^\top,$$

so that we have for a predefined distance $\gamma > 0$ ($\gamma < \alpha$)

$$\|(x(\hat{a}^0 + s^i v^i), y(\hat{a}^0 + s^i v^i)) - (x^0, y^0)\| = \gamma, \quad i = 1, 2,$$

with $(t(\hat{a}), x(\hat{a}), y(\hat{a}))$ minimal solution of problem (7.12) with parameter $\hat{a}$. With the matrices $\hat{M}$, $\hat{N}$ as in Theorem 7.3 we conclude that this aim is approximately fulfilled for

$$|s^i| := \frac{\gamma}{\left\| \left(\hat{M}^{-1} \hat{N} \cdot v^i \right) \big|_{(x,y)} \right\|}, \quad i = 1, 2,$$

with $\left(\hat{M}^{-1} \hat{N} \cdot v^i \right) \big|_{(x,y)}$ the vector consisting of the second to the $(n_1 + 2)$th entry of the vector $\hat{M}^{-1} \hat{N} \cdot v^i$. For a predefined desired number $n^D \in \mathbb{N}$ of new discretization points we set

$$\hat{a} := \hat{a}^0 + l^1 \cdot s^1 v^1 + l^2 \cdot s^2 v^2, \text{ for } l^1, l^2 \in \{-n^D, \ldots, n^D\} \subset \mathcal{Z}, \quad (7.13)$$
$$(l^1, l^2) \neq (0, 0).$$

Solving (7.12) for these new parameters $\hat{a}$ we get minimal solutions (t, x, y) and with that new points of the induced set. We set

$$A^1_{(x^0,y^0)} := \{(x,y) \in \mathbb{R}^{n_1+1} \mid \exists t \in \mathbb{R} \text{ with } (t,x,y) \text{ a minimal solution of}$$
$$(7.12) \text{ for a parameter } \hat{a} \text{ as in } (7.13)\}.$$

By doing this for all points $(x,y) \in \mathcal{M}(F(A^0), \mathbb{R}^2_+)$ we get the following new approximation of the induced set Ω

$$A^1 := A^0 \cup \bigcup_{(x,y) \in \mathcal{M}(F(A^0), \mathbb{R}^2_+)} A^1_{(x,y)}. \tag{7.14}$$

We map these points again under the upper level function F and select only the non-dominated (or ε-EP-minimal) points $\mathcal{M}(F(A^1), \mathbb{R}^2_+)$. Here we can use Lemma 1.11 and consider only the set

$$\tilde{A}^1 := \mathcal{M}(F(A^0), \mathbb{R}^2_+) \cup \bigcup_{(x,y) \in \mathcal{M}(F(A^0), \mathbb{R}^2_+)} A^1_{(x,y)}$$

because we have

$$\mathcal{M}(F(A^1), \mathbb{R}^2_+) = \mathcal{M}(F(\tilde{A}^1), \mathbb{R}^2_+). \tag{7.15}$$

The set $\mathcal{M}(F(A^1), \mathbb{R}^2_+)$ is now an improved approximation of the solution set of the multiobjective bilevel optimization problem, but this set can be refined further by repeating the described steps arbitrarily often.

7.4 Algorithm

We summarize the described steps for the approximation and refinement of the induced set in the following algorithm. The result of this algorithm is an approximation of the solution set of the multiobjective bilevel optimization problem. Because the upper level objective function is vector-valued, too, we get not only one solution but an approximation of the whole efficient set. We assume $K^1 = K^2 = \mathbb{R}^2_+$, $m_1 = m_2 = 2$, $C = \mathbb{R}^p_+$, $n_2 = 1$, $G = \{(x,y) \in \mathbb{R}^{n_1+1} \mid g_j(x,y) \geq 0, j = 1, \ldots, p\}$, $\tilde{G} = [c,d] \subset \mathbb{R}$ for $c,d \in \mathbb{R}$ and we choose $r = (0,1)^\top$ constant.

7 Application to Multiobjective Bilevel Optimization

Algorithm 6 (Multiobjective Bilevel Algorithm).

Step 1: Choose $\beta > 0$ or $n^y \in \mathbb{N}$ and discretize the interval $\tilde{G} = [c, d]$ by

$$y^1 := c, \; y^2 := y^1 + \beta, \; y^3 := y^1 + 2\beta, \ldots, y^{n^y} := y^1 + (n^y - 1)\beta$$

with $n^y \leq d$, $n^y \in \mathbb{N}$.

Step 2: For all $y = y^k$, $k = 1, \ldots, n^y$, determine an approximation $A^{0,k}$ of problem (7.1) with the help of problem $(SP(a, r, \tilde{a}))$ for $\tilde{a} = y^k$. For that give a distance $\alpha > 0$, set $k = 1$ and continue with the following steps:

Step 2a: Solve $f_1(\bar{x}^1) := \min_x \{f_1(x) \mid (x, y^k) \in G\}$ and $f_2(\bar{x}^2) := \min\{f_2(x) \mid (x, y^k) \in G\}$ and determine $a^1 := (f_1(\bar{x}^1), 0)$ and $a^E := (f_1(\bar{x}^2), 0)$. Set $A^{0,k} := \{(\bar{x}^1, y^k)\}$, $a^2 := (f_1(\bar{x}^1) + \delta, 0)^\top$ for a small $\delta > 0$ and $l := 2$.

Step 2b: If $a^l_1 \leq a^E_1$ solve problem $(SP(a^l, r, \tilde{a}))$ for $\tilde{a} = y^k$ with minimal solution x^l and Lagrange multipliers (μ^l, ν^l) and set $A^{0,k} := A^{0,k} \cup \{(x^l, y^k)\}$. Calculate the matrices M, N according to Theorem 3.16, determine $\bar{\lambda}$ according to (7.11) with $\tilde{v} := (1, 0, 0, 0)^\top$, set $a^{l+1} := a^l + \bar{\lambda} \cdot (1, 0)^\top$, $l := l + 1$ and repeat Step 2b.

Else set $k := k + 1$. If $k \leq n^y$ go to Step 2a, else go to Step 3.

Step 3: Set $A^0 := \bigcup_{k=1}^{n^y} A^{0,k}$ and determine $\mathcal{M}(F(A^0), \mathbb{R}^2_+)$ with the help of a Pareto filter or the Jahn-Graef-Younes method. Set $i := 0$ and choose $\gamma^0 > 0$.

Step 4: For any point $(x, y) \in \mathcal{M}(F(A^i), \mathbb{R}^2_+)$ determine a refinement of the induced set around this point by solving the scalarization (7.12) and by choosing the parameters $\hat{a} \in \mathbb{R}^3$ as in (7.13). Determine A^{i+1} by (7.14).

Step 5: Calculate $\mathcal{M}(F(A^{i+1}), \mathbb{R}^2_+)$ by using (7.15) and a Pareto filter or the Jahn-Graef-Younes method. If this approximation of the solution set of the multiobjective bilevel problem is sufficient, then stop.

Else set $i := i + 1$, choose $\gamma^i > 0$ and go to Step 4.

With the approximations $A^{0,k}$ determined in the Steps 2a and 2b an approximation of the solution set of the tricriteria optimization problem (7.4) and hence of the induced set Ω is calculated. In Step 2a it is necessary to choose a^2 as described to avoid numerical difficulties. The set $\mathcal{M}(F(A^i), \mathbb{R}^2_+)$ is the searched approximation of the solution set of the multiobjective bilevel optimization problem.

For Algorithm 6 we have assumed $n_2 = 1$, i.e. the compact set $\tilde{G} = [c, d]$ is an interval. If it is $n_2 \geq 2$ an (equidistant) discretization of the compact set $\tilde{G}$ has to be found. This can be done e.g. by finding a cuboid $I := [c_1, d_1] \times \ldots \times [c_{n_2}, d_{n_2}] \subset \mathbb{R}^{n_2}$ with $\tilde{G} \subset I$ and discretizing this cuboid with points

$$\{y \in \mathbb{R}^{n_2} \mid y_i = c_i + n_i \beta_i \text{ for a } n_i \in \mathbb{N} \cup \{0\}, \ y_i \leq d_i, \ i = 1, \ldots, n_2\}$$

for distances $\beta_i > 0$ ($i = 1, \ldots, n_2$) and then selecting only the discretization points being an element of $\tilde{G}$.

7.5 Numerical Results

In this section we apply the proposed algorithm to a nonlinear academic test problem for illustrating the single steps of Algorithm 6. Further we discuss a nonlinear problem which arose in a technical application in medical engineering.

7.5.1 Test Problem

We consider the following multiobjective bilevel problem with $n_1 = 2$, $n_2 = 1$, $m_1 = m_2 = 2$ and $K^1 = K^2 = \mathbb{R}^2_+$.

… 7 Application to Multiobjective Bilevel Optimization

$$\min_{y} \begin{pmatrix} F_1(x,y) \\ F_2(x,y) \end{pmatrix} := \begin{pmatrix} x_1 + x_2^2 + y + \sin^2(x_1 + y) \\ \cos(x_2) \cdot (0.1 + y) \cdot (\exp(-\frac{x_1}{0.1 + x_2})) \end{pmatrix}$$

subject to the constraints

$$x \in \operatorname{argmin}_x \left\{ \begin{pmatrix} f_1(x,y) \\ f_2(x,y) \end{pmatrix} \;\middle|\; (x,y) \in G \right\},$$

$$y \in [0, 10]$$

with $f_1, f_2 \colon \mathbb{R}^3 \to \mathbb{R}$,

$$f_1(x,y) := \frac{(x_1 - 2)^2 + (x_2 - 1)^2}{4} + \frac{x_2 y + (5 - y)^2}{16} + \sin\left(\frac{x_2}{10}\right),$$

$$f_2(x,y) := \frac{x_1^2 + (x_2 - 6)^4 - 2x_1 y - (5 - y)^2}{80}$$

and

$$G := \{(x_1, x_2, y) \in \mathbb{R}^3 \mid x_1^2 - x_2 \leq 0,\; 5x_1^2 + x_2 \leq 10,$$
$$x_2 - (5 - y/6) \leq 0,\; x_1 \geq 0\}.$$

We have chosen $n_1 = 2$, i.e. $x \in \mathbb{R}^2$, for being able to plot the approximation of the induced set $\Omega \subset \mathbb{R}^3$. We use the optimistic approach and hence solve

$$\min_{x,y} \begin{pmatrix} F_1(x,y) \\ F_2(x,y) \end{pmatrix}$$

subject to the constraints

$$(x, y) \in \Omega.$$

The induced set $\Omega \subset \mathbb{R}^3$ is according to Theorem 7.1 equal to the solution set of the tricriteria optimization problem

$$\min_{x,y} \begin{pmatrix} f_1(x,y) \\ f_2(x,y) \\ y \end{pmatrix}$$

subject to the constraints

$$(x, y) \in G,$$
$$y \in [0, 10]$$

w. r. t. the ordering cone $\hat{K} = \mathbb{R}_+^2 \times \{0\}$.

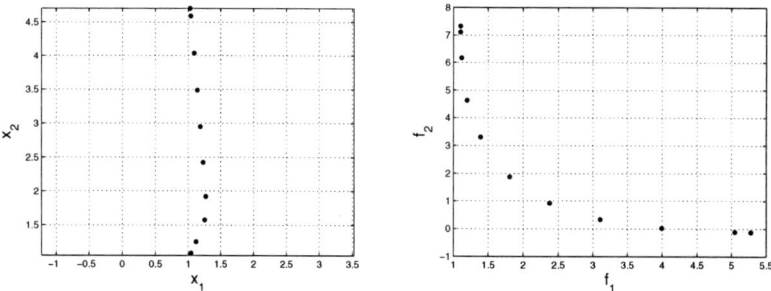

Fig. 7.1. Approximation (a) of the solution set and (b) of the efficient set of the problem (7.1) for $y = 1.8$ for the test problem.

We apply Algorithm 6 with $\beta = 0.6$ for discretizing the interval $[0, 10]$ in Step 1. For example for $y = 1.8$ and the distance $\alpha = 0.6$ the Steps 2a and 2b lead to the approximation of the minimal solution set of the problem (7.1) shown in Fig. 7.1,a) and with that to an approximation of the efficient set of the problem (7.1) shown in Fig. 7.1,b). Repeating the Steps 2a and 2b for all discretization points y of the interval $[0, 10]$ we get the set A^0 shown in Fig. 7.2,a) which is an approximation of the $\hat{K}$-minimal set of the problem (7.4) and hence of the induced set. In Fig. 7.2,b) the result of Step 3, the set $F(A^0)$, is drawn. Here the image points under F of the set $\mathcal{M}(F(A^0), \mathbb{R}^2_+)$ are marked with circles and are connected with lines.

We now start the refinement and continue the algorithm by choosing $\gamma^0 = 0.3$ and $n^D = 2$. The set A^1, the refinement of the set A^0 according to Step 4, is given in Fig. 7.3,a), whereby the points $A^1 \setminus A^0$ are drawn in black and the points A^0 in gray. The set $F(A^1)$ and the non-dominated points of this set can be seen in Fig. 7.3,b), compare Step 5 of the algorithm. Repeating this with $\gamma^1 = 0.21$ and $\gamma^2 = 0.12$ and doing the refinement in Step 4 only for those points of the set $\mathcal{M}(F(A^i), \mathbb{R}^2_+)$ which have no neighbors in the set $F(\mathcal{M}(F(A^i), \mathbb{R}^2_+))$ with a distance less than 0.3, we get the results presented in Figures 7.4,a), b) and 7.5,a), b).

Then the algorithm is stopped as only small improvements of the approximation of the efficient set of the multiobjective bilevel optimization problem were gained by the last iteration and as the distances between the points were accepted as sufficiently small.

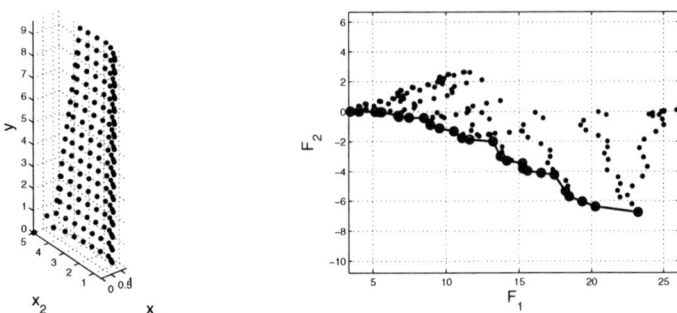

Fig. 7.2. (a) Approximation A^0 of induced set Ω and (b) the image $F(A^0)$.

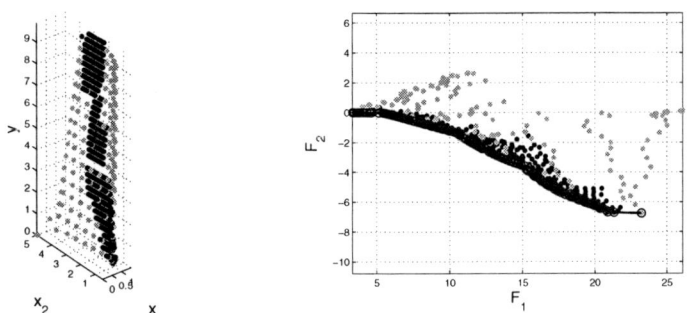

Fig. 7.3. (a) Refinement A^1 of the induced set Ω and (b) the image set $F(A^1)$.

7.5.2 Application Problem

The following multiobjective bilevel optimization problem arose during the examination of a recent problem in medical engineering ([184, pp.62f]) dealing with the configuration of coils. In its original version this problem is a usual scalar-valued standard optimization problem which had to be reformulated as a bilevel optimization problem due to the need of real-time solutions and because of its structure:

$$\min_y \|x(y)\|_2^2$$
$$\text{subject to the constraints}$$
$$x(y) \in \mathrm{argmin}_x\{\|x\|_2^2 \mid A(y) \cdot Vx = b(y),\ x \in \mathbb{R}^{14}\},$$
$$y \in [0, \pi]. \tag{7.16}$$

The vector $b(y) \in \mathbb{R}^6$ is given by

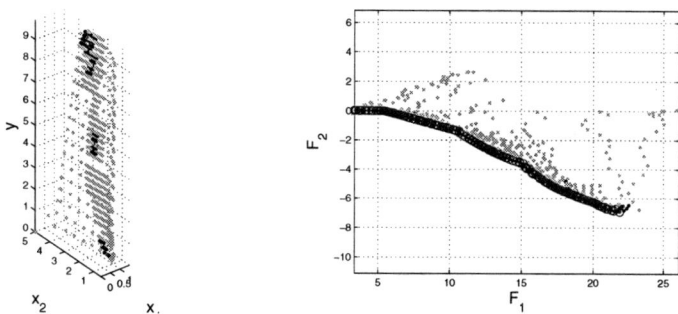

Fig. 7.4. (a) The set A^2 and (b) the image set $F(A^2)$.

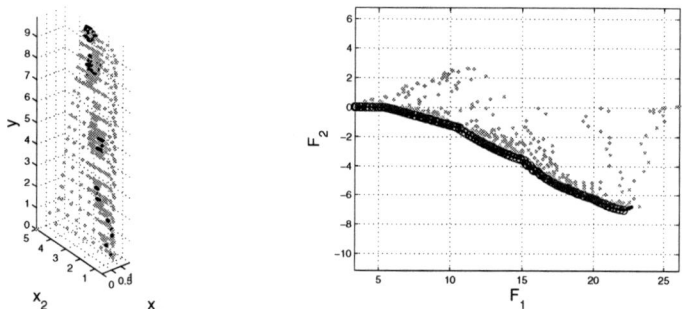

Fig. 7.5. (a) The set A^3 and (b) the image set $F(A^3)$.

$$b(y) := (m_x(y), m_y(y), m_z(y), m_F)^\top$$

with $m_F \in \mathbb{R}^3$ arbitrarily and $m_x, m_y, m_z \colon \mathbb{R} \to \mathbb{R}$ defined by

$$\begin{pmatrix} m_x(y) \\ m_y(y) \\ m_z(y) \end{pmatrix} := m_0(\tau_1 \cos y + \tau_2 \sin y)$$

for a scalar $m_0 \in \mathbb{R}$, $\tau_1, \tau_2 \in \mathbb{R}^3$ with

$$\tau_1^\top m_F = 0, \qquad \tau_2^\top m_F = 0, \qquad \|\tau_1\|_2 = \|\tau_2\|_2 = 1.$$

In the following we choose

$$m_F := \begin{pmatrix} 1 \\ 0 \\ 0 \end{pmatrix}, \quad \tau_1 := \begin{pmatrix} 0 \\ 1 \\ 0 \end{pmatrix}, \quad \tau_2 := \begin{pmatrix} 0 \\ 0 \\ 1 \end{pmatrix}, \quad m_0 := 1$$

204   7 Application to Multiobjective Bilevel Optimization

and hence we get

$$b(y) = (0, \cos y, \sin y, 1, 0, 0)^\top.$$

The matrix $A(y) \in \mathbb{R}^{6 \times 8}$ is given by

$$A(y) := \begin{pmatrix} 1 & 0 & 0 & 0 & 0 & 0 & 0 & 0 \\ 0 & 1 & 0 & 0 & 0 & 0 & 0 & 0 \\ 0 & 0 & 1 & 0 & 0 & 0 & 0 & 0 \\ 0 & 0 & 0 & m_x(y) & m_y(y) & m_z(y) & 0 & 0 \\ 0 & 0 & 0 & 0 & m_x(y) & 0 & m_z(y) & m_y(y) \\ 0 & 0 & 0 & -m_z(y) & 0 & m_x(y) & m_y(y) & -m_z(y) \end{pmatrix}$$

$$= \begin{pmatrix} 1 & 0 & 0 & 0 & 0 & 0 & 0 & 0 \\ 0 & 1 & 0 & 0 & 0 & 0 & 0 & 0 \\ 0 & 0 & 1 & 0 & 0 & 0 & 0 & 0 \\ 0 & 0 & 0 & 0 & \cos y & \sin y & 0 & 0 \\ 0 & 0 & 0 & 0 & 0 & 0 & \sin y & \cos y \\ 0 & 0 & 0 & -\sin y & 0 & 0 & \cos y & -\sin y \end{pmatrix}.$$

The matrix $V \in \mathbb{R}^{8 \times 14}$ is a non-sparse matrix with $\text{rank}(V) = 8$ which depends on the considered medical system. For the calculation a randomly chosen matrix is taken here. The matrix V is (after rounding) equal to

$$\begin{pmatrix} 0.9501 & 0.8214 & 0.9355 & 0.1389 & 0.4451 & 0.8381 & 0.3046 & 0.3784 & 0.8180 & 0.8385 & 0.7948 & 0.8757 & 0.2844 & 0.4329 \\ 0.2311 & 0.4447 & 0.9169 & 0.2028 & 0.9318 & 0.0196 & 0.1897 & 0.8600 & 0.6602 & 0.5681 & 0.9568 & 0.7373 & 0.4692 & 0.2259 \\ 0.6068 & 0.6154 & 0.4103 & 0.1987 & 0.4660 & 0.6813 & 0.1934 & 0.8537 & 0.3420 & 0.3704 & 0.5226 & 0.1365 & 0.0648 & 0.5798 \\ 0.4860 & 0.7919 & 0.8936 & 0.6038 & 0.4186 & 0.3795 & 0.6822 & 0.5936 & 0.2897 & 0.7027 & 0.8801 & 0.0118 & 0.9883 & 0.7604 \\ 0.8913 & 0.9218 & 0.0579 & 0.2722 & 0.8462 & 0.8318 & 0.3028 & 0.4966 & 0.3412 & 0.5466 & 0.1730 & 0.8939 & 0.5828 & 0.5298 \\ 0.7621 & 0.7382 & 0.3529 & 0.1988 & 0.5252 & 0.5028 & 0.5417 & 0.8998 & 0.5341 & 0.4449 & 0.9797 & 0.1991 & 0.4235 & 0.6405 \\ 0.4565 & 0.1763 & 0.8132 & 0.0153 & 0.2026 & 0.7095 & 0.1509 & 0.8216 & 0.7271 & 0.6946 & 0.2714 & 0.2987 & 0.5155 & 0.2091 \\ 0.0185 & 0.4057 & 0.0099 & 0.7468 & 0.6721 & 0.4289 & 0.6979 & 0.6449 & 0.3093 & 0.6213 & 0.2523 & 0.6614 & 0.3340 & 0.3798 \end{pmatrix}.$$

In the medical-engineering system further objectives are of interest, which had not yet been examined numerically. For example it is important that the result $x(y)$ for a new system adjustment, expressed by a changed matrix V, does not differ too much from the prior minimal solution $x_{\text{old}} \in \mathbb{R}^{14}$. Thus, as additional objective function of the upper level, we are interested in the objective function

$$\|x(y) - x_{\text{old}}\|_2^2 \to \min!$$

7.5 Numerical Results

We set

$$x_{\text{old}} := (0.1247, 0.1335, -0.0762, -0.1690, 0.2118, -0.0534, -0.1473,$$
$$0.3170, -0.0185, -0.1800, 0.1700, -0.0718, 0.0058, 0.0985)^\top.$$

Besides, the constraint $A(y) \cdot Vx = b(y)$ is only a soft constraint and a not too strict fulfilling of this constraint in favor of a smaller objective function value on the lower level is acceptable. Because of that we extend the lower level optimization problem by the objective function

$$\|A(y) \cdot Vx - b(y)\|_2^2 \to \min!$$

The violation of this former constraint is bounded by the constraint $\|A(y) \cdot Vx - b(y)\|_2^2 \leq \Delta_{\max}$. We choose $\Delta_{\max} := 0.3$. We get the following multiobjective bilevel optimization problem:

$$\min_y \begin{pmatrix} F_1(x,y) \\ F_2(x,y) \end{pmatrix} = \begin{pmatrix} \|x(y)\|_2^2 \\ \|x(y) - x_{\text{old}}\|_2^2 \end{pmatrix}$$

subject to the constraints

$$x = x(y) \in \operatorname{argmin}_x \left\{ \begin{pmatrix} f_1(x,y) \\ f_2(x,y) \end{pmatrix} = \begin{pmatrix} \|A(y) \cdot Vx - b(y)\|_2^2 \\ \|x\|_2^2 \end{pmatrix} \right|$$
$$\|A(y) \cdot Vx - b(y)\|_2^2 \leq \Delta_{\max}, x \in \mathbb{R}^{14} \bigg\},$$
$$y \in [0, \pi].$$
(7.17)

For solving this problem we use the optimistic approach. The constraint set of the upper level problem is equal to the set of minimal solutions of the following tricriteria optimization problem

$$\min_{x,y} \begin{pmatrix} \|A(y) \cdot Vx - b(y)\|_2^2 \\ \|x\|_2^2 \\ y \end{pmatrix}$$

subject to the constraints (7.18)

$$\|A(y) \cdot Vx - b(y)\|_2^2 \leq \Delta_{\max},$$
$$x \in \mathbb{R}^{14},$$
$$y \in [0, \pi]$$

7 Application to Multiobjective Bilevel Optimization

w. r. t. the ordering cone $\mathbb{R}^2_+ \times \{0\}$. For solving this problem we start according to Step 1 in the Algorithm 6 with a discretization of the interval $[0, \pi]$ by choosing $\beta = \pi/8$:

$$0.0001,\ 0.0001 + \frac{\pi}{8},\ 0.0001 + 2 \cdot \frac{\pi}{8},\ 0.0001 + 3 \cdot \frac{\pi}{8},\ \ldots,\ \pi.$$

We do not start with $y^1 = 0$ for numerical reasons.

For $y^k \in \{0.0001,\ 0.0001 + \frac{\pi}{8},\ 0.0001 + 2 \cdot \frac{\pi}{8}, \ldots, \pi\}$ we then determine an approximation of the set of EP-minimal points of the bicriteria optimization problem

$$\min_{x} \begin{pmatrix} \|A(y^k) \cdot Vx - b(y^k)\|_2^2 \\ \|x\|_2^2 \end{pmatrix}$$

subject to the constraints

$$\|A(y^k) \cdot Vx - b(y^k)\|_2^2 \leq \Delta_{\max},$$
$$x \in \mathbb{R}^{14}.$$

For solving this problem we use the scalarization according to $(\text{SP}(a, r, \tilde{a}))$ with $a = (a_1, 0)^\top$ and $r = (0, 1)^\top$. Hence we solve

$$\min_{t,x} t$$

subject to the constraints
$$a_1 - \|A(y^k) \cdot Vx - b(y^k)\|_2^2 \geq 0,$$
$$t - \|x\|_2^2 \geq 0, \tag{7.19}$$
$$\|A(y) \cdot Vx - b(y)\|_2^2 \leq \Delta_{\max},$$
$$x \in \mathbb{R}^{14}.$$

For $a_1 < 0$ the constraint set of the problem (7.19) is empty. Due to the constraint $\|A(y) \cdot Vx - b(y)\|_2^2 \leq \Delta_{\max}$ solving the problem (7.19) for $a_1 > \Delta_{\max}$ is equal to solve it for $a_1 = \Delta_{\max}$. Hence it is sufficient to consider the following problem for parameters $a \in H = \{y \in \mathbb{R}^2 \mid y_2 = 0\}$ with $a_1 \in [0, \Delta_{\max}]$:

$$\min_{x} \|x\|_2^2$$

subject to the constraints $\tag{7.20}$
$$\|A(y^k) \cdot Vx - b(y^k)\|_2^2 \leq a_1,$$
$$x \in \mathbb{R}^{14}.$$

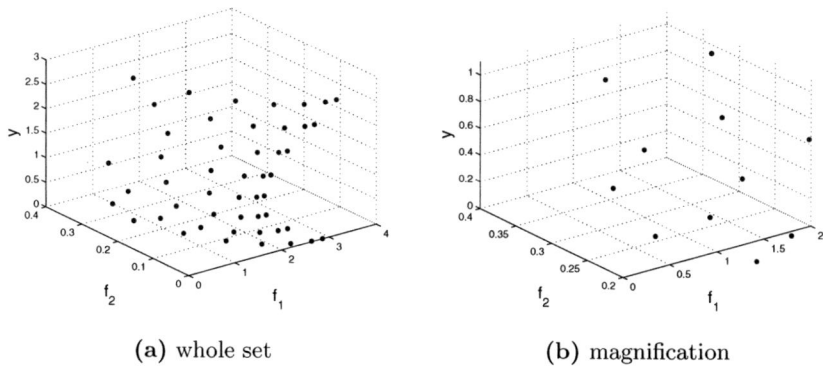

(a) whole set   (b) magnification

Fig. 7.6. Approximation of the efficient set of the tricriteria optimization problem (7.18).

We vary the parameter a_1 so that the approximation points have nearly a distance of $\alpha = 0.2$ (Step 2). Summarizing the result for all discretization points $y \in [0, \pi]$ delivers a first approximation A^0 of the induced set Ω. This set is, as a subset of $\mathbb{R}^{15}$, no longer visualizable. However, as the induced set is at the same time the solution set of the tricriteria optimization problem (7.18) we present in Fig. 7.6,a) the image of the approximation under the objective functions of the tricriteria problem. In the magnification Fig. 7.6,b) the most interesting part is shown.

According to Step 3 we map the approximation A^0 of the induced set with the vector-valued objective function F of the upper level problem and get the set $F(A^0)$ plotted in Fig. 7.7,a). The nondominated points are marked with circles and connected with lines. In Fig. 7.7,b) the interesting part is scaled up. Pursuant Step 4 of the algorithm we refine the induced set in a neighborhood of the three nondominated points with the results given in the Figures 7.8,a), b) and 7.9,a), b). We have chosen $\gamma^0 = \alpha/7 = 1/35$ and $n^D = 3$.

A second refinement with distance $\gamma^1 = 1/50$ results in the points in the Figures 7.10 and 7.11. We conclude with a third refinement step with $\gamma^2 = 1/70$. The final results are shown in the Figures 7.12 and 7.13.

We examine the non-dominated points $\mathcal{M}(F(A^2), \mathbb{R}^2_+) \subset \mathbb{R}^{15}$ after the third refinement. This set is an approximation of the solution set of the multiobjective bilevel problem. We observe, that for all $(x, y) \in \mathcal{M}(F(A^2), \mathbb{R}^2_+)$ we have $y \in [0.6838, 0.7958] \subset [0, \pi]$ as well

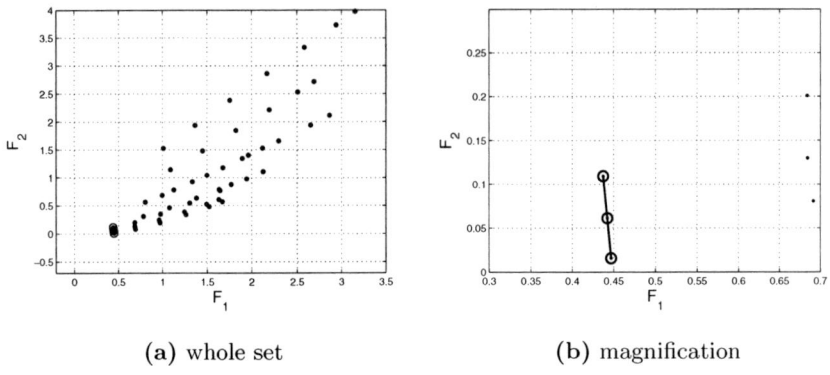

Fig. 7.7. Image $F(A^0)$ of the approximation A^0 of the induced set.

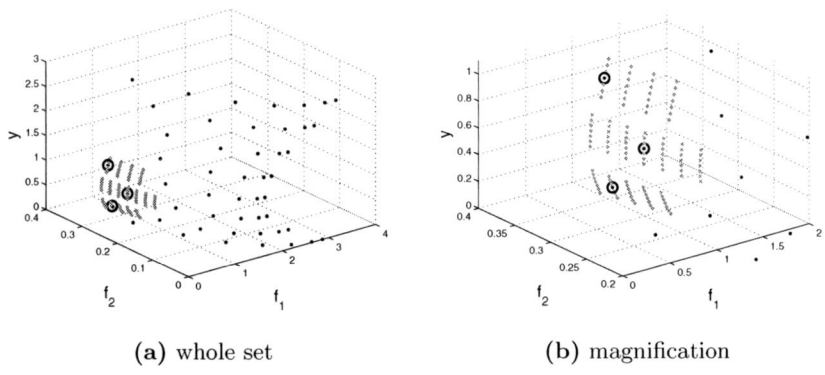

Fig. 7.8. Efficient set of the tricriteria optimization problem (7.18) after the first refinement.

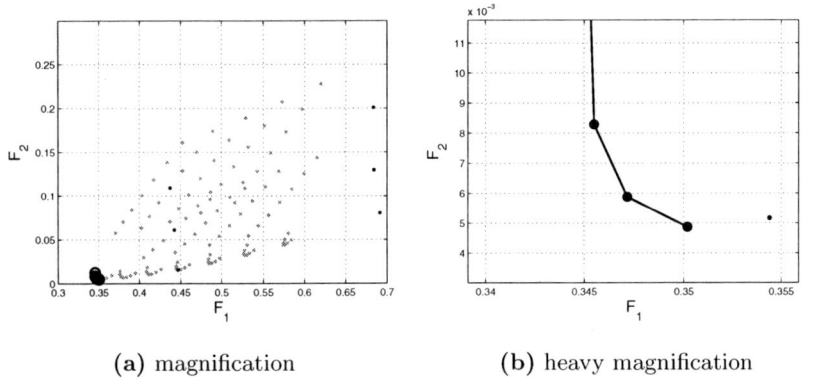

Fig. 7.9. Image $F(A^1)$ of the refined approximation A^1 of the induced set.

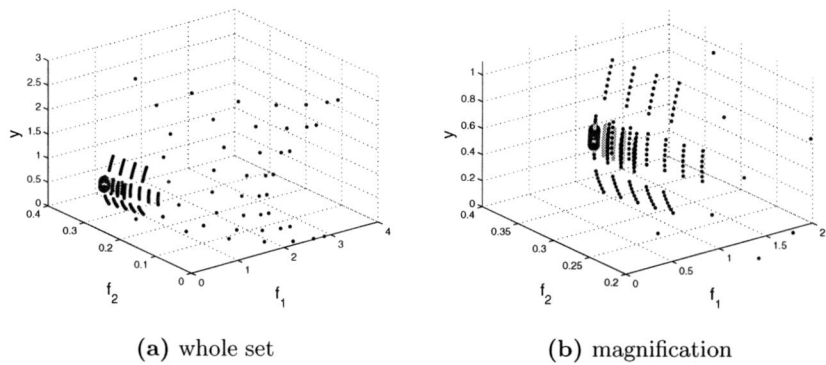

Fig. 7.10. Efficient set of the tricriteria optimization problem (7.18) after the second refinement.

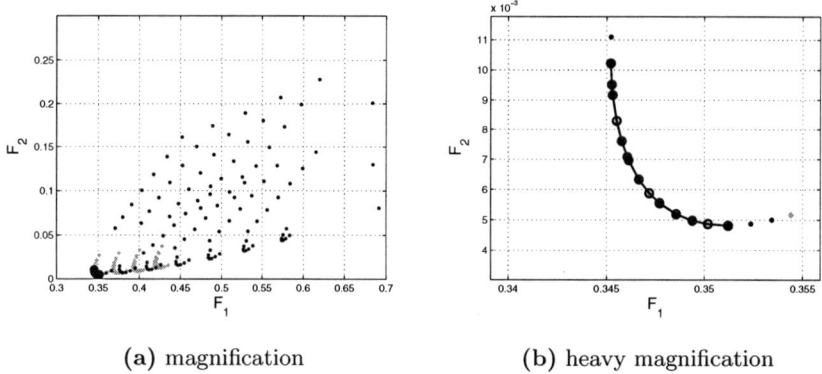

Fig. 7.11. Image $F(A^2)$ of the twice refined approximation A^2 of the induced set.

as $x_1, x_2, x_5, x_8, x_{11}, x_{13}, x_{14} > 0$ and $x_3, x_4, x_6, x_7, x_9, x_{10}, x_{12} < 0$. The values of the points $(x, y) \in \mathcal{M}(F(A^2), \mathbb{R}_+^2)$ under the functions F_1, F_2 and f_1, f_2, respectively, range between the following intervals:

$$F_1(x, y) = \|x\|_2^2 \qquad \in [0.3452, 0.3512],$$
$$F_2(x, y) = \|x - x_{alt}\|_2^2 \qquad \in [0.0048, 0.0105],$$
$$f_1(x, y) = \|x\|_2^2 \qquad \in [0.3452, 0.3512],$$
$$f_2(x, y) = \|A(y) \cdot Vx - b(y)\|_2^2 = 0.2897.$$

Hence the values of $f_2(x, y)$ are constant for all $(x, y) \in \mathcal{M}(F(A^2), \mathbb{R}_+^2)$ and thus only those points of the discretized constraint set are non-dominated which are generated by solving the problem (7.20) for

210   7 Application to Multiobjective Bilevel Optimization

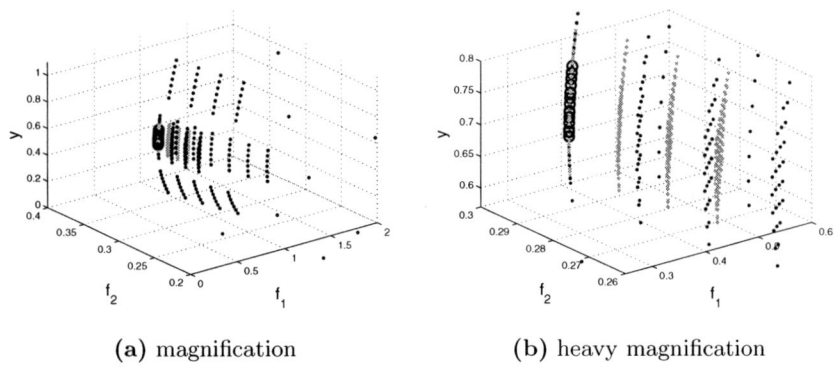

Fig. 7.12. Efficient set of the tricriteria optimization problem (7.18) after the third refinement.

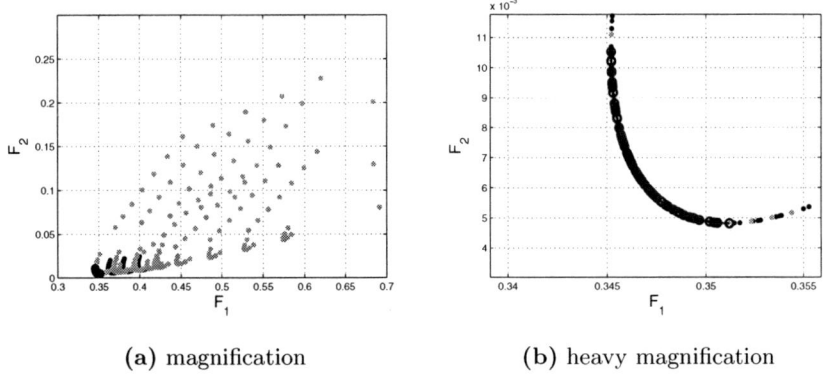

Fig. 7.13. Image $F(A^0)$ of the three times refined approximation A^0 of the induced set.

the parameter $a_1 = 0.2897$. The d.m. of the upper level, the so-called leader, can now choose his favorable solution from the set $\mathcal{M}(F(A^2), \mathbb{R}^2_+)$.

7.6 Multiobjective Bilevel Optimization Problems with Coupled Upper Level Constraints

In this section we consider instead of problem (7.3) the more general formulation

7.6 Coupled Upper Level Constraints

$$\min_{x,y} F(x,y)$$
subject to the constraints
$$x \in M_y(f(G), K^1),$$
$$(x,y) \in \tilde{G}$$
(7.21)

with the constraint set $\tilde{G} \subset \mathbb{R}^{n_1} \times \mathbb{R}^{n_2}$. This results in a coupling of the upper level variable y and the lower level variable x. Then

$$\Omega' := \{(x,y) \in \mathbb{R}^{n_1} \times \mathbb{R}^{n_2} \mid x \in M_y(f(G), K^1), \ (x,y) \in \tilde{G}\}$$

denotes the induced set of the problem (7.21).

First notice that the constraint $(x,y) \in \tilde{G}$ from the upper level cannot just be moved to the lower level as the following example demonstrates.

Example 7.4. We consider the bilevel optimization problem

$$\min_{x,y} y + \tfrac{1}{2}x$$
subject to the constraints
$$x \in \operatorname{argmin}_x\{x \in \mathbb{R} \mid x \in [0,2], \ x \geq y\},$$
$$x \geq 2 - y,$$
$$y \in [0,2].$$

The lower level minimal solution dependent on $y \in [0,2]$ is thus $x(y) = y$ and hence only the points $(x(y), y) = (y, y)$ for $y \in [1,2]$ are feasible for the upper level. This leads to the optimal value $3/2$ of the upper level function and the minimal solution $(x,y) = (1,1)$.

Moving the constraint $x \geq 2 - y$ instead to the lower level, i.e. considering the problem

$$\min_x\{x \mid x \in [0,2], \ x \geq y, \ x \geq 2 - y\}$$

on the lower level leads to the lower level minimal solutions

$$x(y) = \begin{cases} 2 - y, & \text{for } y \in [0,1], \\ y, & \text{for } y \in]1,2]. \end{cases}$$

Then the upper level optimal value is 1 with minimal solution $(x,y) = (2,0)$.

7 Application to Multiobjective Bilevel Optimization

This has of course also a practical interpretation. The same constraint on the lower level restricting the constraint set on the lower level has a different meaning as on the upper level. There, the feasibility is restricted after the determination of the minimal solution of the lower level and is thus an implicit constraint. For more details see [49, pp.25f].

We first try to generalize the results of Sect. 7.2 and thus we consider the multiobjective optimization problem

$$\min_{x,y} \hat{f}(x,y) := \begin{pmatrix} f(x,y) \\ y \end{pmatrix}$$

subject to the constraints

$$(x,y) \in G, \ (x,y) \in \tilde{G}$$

(7.22)

w. r. t. the ordering cone $\hat{K} := K^1 \times \{0_{n_2}\} \subset \mathbb{R}^{m_1} \times \mathbb{R}^{n_2}$. Let $\hat{G} := G \cap \tilde{G}$ denote the constraint set of the problem (7.22). Then the $\hat{K}$-minimal solution set of (7.22) is denoted as

$$\hat{\mathcal{M}}' := \mathcal{M}(\hat{f}(\hat{G}), \hat{K}).$$

This set $\hat{\mathcal{M}}'$ has a close connection to the induced set Ω' as the following theorem shows. However we get no equivalence result like in the uncoupled case.

Theorem 7.5. *Let $\hat{\mathcal{M}}'$ be the set of $\hat{K}$-minimal points of the multiobjective optimization problem (7.22) with $\hat{K} = K^1 \times \{0_{n_2}\}$. Then $\Omega' \subset \hat{\mathcal{M}}'$.*

Proof. We have the equivalences

$(\bar{x}, \bar{y}) \in \Omega' \Leftrightarrow \bar{x} \in G(\bar{y}) \wedge (\bar{x}, \bar{y}) \in \tilde{G} \wedge$
  $(\nexists x \in G(\bar{y}) \text{ with } f(\bar{x}, \bar{y}) \in f(x, \bar{y}) + K^1 \setminus \{0_{m_1}\})$
$\Rightarrow \bar{x} \in G(\bar{y}) \wedge (\bar{x}, \bar{y}) \in \tilde{G} \wedge$
  $(\nexists x \in G(\bar{y}) \text{ with } (x, \bar{y}) \in \tilde{G} \text{ and }$
  $f(\bar{x}, \bar{y}) \in f(x, \bar{y}) + K^1 \setminus \{0_{m_1}\})$
$\Leftrightarrow (\bar{x}, \bar{y}) \in G \wedge (\bar{x}, \bar{y}) \in \tilde{G} \wedge$
  $(\nexists (x, y) \in G \cap \tilde{G} \text{ with } y = \bar{y} \text{ and }$
  $f(\bar{x}, \bar{y}) \in f(x, y) + K^1 \setminus \{0_{m_1}\})$
$\Leftrightarrow (\bar{x}, \bar{y}) \in \hat{G} \wedge$
  $(\nexists (x, y) \in \hat{G} \text{ with } \hat{f}(\bar{x}, \bar{y}) \in \hat{f}(x, y) + \hat{K} \setminus \{0_{m_1+n_2}\})$
$\Leftrightarrow (\bar{x}, \bar{y}) \in \hat{\mathcal{M}}'.$ □

7.6 Coupled Upper Level Constraints

In contradiction to Sect. 7.2 we do not have $\hat{\mathcal{M}}' \subset \Omega'$ generally because it can happen for a point $(\bar{x}, \bar{y}) \in \hat{\mathcal{M}}'$ that there is no x with $(x, \bar{y}) \in \hat{G}$ and with

$$f(\bar{x}, \bar{y}) \in f(x, \bar{y}) + K^1 \setminus \{0_{m_1}\}, \qquad (7.23)$$

but there exists a point x with $(x, \bar{y}) \in G$, $(x, \bar{y}) \notin \tilde{G}$ and (7.23). This is demonstrated in the following example.

Example 7.6. We consider the biobjective bilevel optimization problem

$$\min_y F(x, y) = \begin{pmatrix} x_1 - y \\ x_2 \end{pmatrix}$$

subject to the constraints

$$x = x(y) \in \operatorname{argmin}_x \left\{ f(x, y) = \begin{pmatrix} x_1 \\ x_2 \end{pmatrix} \,\middle|\, (x, y) \in G \right\},$$

$$(x, y) \in \tilde{G}$$

with $n_1 = 2$, $n_2 = 1$, $K^1 = K^2 = \mathbb{R}_+^2$, $m_1 = m_2 = 2$,

$$G := \{(x, y) \in \mathbb{R}^3 \mid \|x\|_2^2 \leq y^2\}$$

and

$$\tilde{G} := \{(x, y) \in \mathbb{R}^3 \mid 0 \leq y \leq 1,\ x_1 + x_2 \geq -1\}.$$

Then

$$\mathcal{M}_y(f(G), K^1) = \{x = (x_1, x_2) \in \mathbb{R}^2 \mid \|x\|_2^2 = y^2,\ x_1 \leq 0,\ x_2 \leq 0\}$$

and thus

$$\Omega' = \{(x, y) \in \mathbb{R}^3 \mid \|x\|_2^2 = y^2,\ x_1 \leq 0,\ x_2 \leq 0,\ 0 \leq y \leq 1,\ x_1 + x_2 \geq -1\}.$$

Let $\hat{\mathcal{M}}'$ be the set of $(\mathbb{R}_+^2 \times \{0\})$-minimal points of the tricriteria optimization problem

$$\min_{x,y} \begin{pmatrix} x_1 \\ x_2 \\ y \end{pmatrix}$$

subject to the constraints

$$(x, y) \in G \cap \tilde{G} = \{(x, y) \in \mathbb{R}^3 \mid 0 \leq y \leq 1,\ x_1 + x_2 \geq -1,\ \|x\|_2^2 \leq y^2\}$$

according to problem (7.22). Then for $(\bar{x}, \bar{y}) = (-\frac{1}{2}, -\frac{1}{2}, 1)^\top$ we have $(\bar{x}, \bar{y}) \in \hat{\mathcal{M}}'$ because $(\bar{x}, \bar{y}) \in G \cap \tilde{G}$ and there is no $(x, y) \in G \cap \tilde{G}$ with

$$\hat{f}(\bar{x}, \bar{y}) \in \hat{f}(x, y) + \hat{K} \setminus \{0_3\}$$

$$\Leftrightarrow \begin{pmatrix} \bar{x}_1 \\ \bar{x}_2 \\ \bar{y} \end{pmatrix} \in \begin{pmatrix} x_1 \\ x_2 \\ y \end{pmatrix} + (\mathbb{R}_+^2 \times \{0\}) \setminus \{0_3\}$$

$$\Leftrightarrow \begin{pmatrix} \bar{x}_1 \\ \bar{x}_2 \end{pmatrix} \in \begin{pmatrix} x_1 \\ x_2 \end{pmatrix} + \mathbb{R}_+^2 \setminus \{0_2\} \wedge y = \bar{y}.$$

The set $\{x \in \mathbb{R}^2 \mid (x, y) \in G \cap \tilde{G}, \ y = 1\}$ is drawn in Fig. 7.14. However $(\bar{x}, \bar{y}) \notin \Omega'$ because $\|\bar{x}\|_2^2 = \frac{1}{2} \neq \bar{y}^2$.

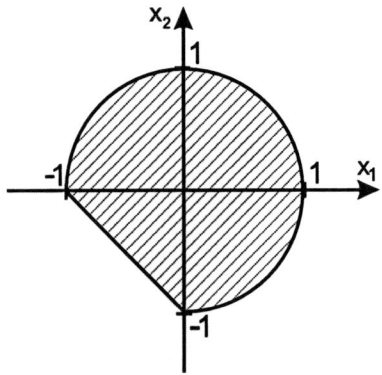

Fig. 7.14. The set $\{x \in \mathbb{R}^2 \mid (x, y) \in G \cap \tilde{G}, \ y = 1\}$ of Example 7.6.

In this example, as the induced set can be determined explicitly, the minimal solution set can be calculated by solving the biobjective optimization problem $\min_{x,y}\{F(x, y) \mid (x, y) \in \Omega'\}$. We get as solution set of the multiobjective bilevel optimization problem:

$$\mathcal{S}^{\min} = \left\{ (x_1, x_2, y) \ \middle| \ x_1 = -1 - x_2, \ x_2 = -\frac{1}{2} \pm \frac{1}{4}\sqrt{8y^2 - 4}, \ y \in \left[\frac{\sqrt{2}}{2}, 1\right] \right\}.$$

The image $\{F(x_1, x_2, y) \mid (x_1, x_2, y) \in \mathcal{S}^{\min}\}$ of this set is

7.6 Coupled Upper Level Constraints

$$\left\{(z_1, z_2) \in \mathbb{R}^2 \;\middle|\; z_1 = -1 - z_2 - y,\; z_2 = -\tfrac{1}{2} \pm \tfrac{1}{4}\sqrt{8y^2 - 4},\right.$$
$$\left. y \in \left[\tfrac{\sqrt{2}}{2}, 1\right]\right\}.$$

These sets are plotted in Fig. 7.15.

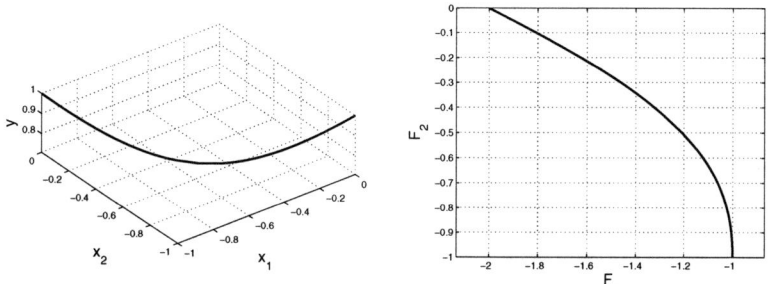

Fig. 7.15. Solution set $\mathcal{S}^{\min}$ of the biobjective bilevel optimization problem of Example 7.6 and the image $F(\mathcal{S}^{\min})$.

Problem (7.22) has the same structure if we formulate it to the bilevel optimization problem

$$\min_{x,y} F(x,y)$$
subject to the constraints (7.24)
$$x \in \mathcal{M}_y(f(G \cap \tilde{G}), K^1).$$

As here no upper level constraint exists we have according to Theorem 7.1 that the $\hat{K}$-minimal solution set $\hat{\mathcal{M}}'$ of (7.22) equals the induced set Ω' of (7.24). Hence, under the conditions of Theorem 7.5 we can of course not gain equality in general as otherwise this would imply that the induced sets of (7.21) and (7.24) are equal. This would mean that it makes no difference on which level the constraint $(x,y) \in \tilde{G}$ is given in contradiction to the Example 7.4.

Theorem 7.5 does not produce an equivalent formulation for the induced set Ω'. This can be reached by considering the following modified multiobjective optimization problem instead of problem (7.22):

7 Application to Multiobjective Bilevel Optimization

$$\min_{x,y} \hat{f}(x,y) = \begin{pmatrix} f(x,y) \\ y \end{pmatrix}$$

subject to the constraints (7.25)

$$(x,y) \in G,$$
$$y \in \tilde{G}^y$$

w. r. t. the ordering cone $\hat{K}$. Here let the compact set $\tilde{G}^y \subset \mathbb{R}^{n_2}$ be an arbitrary compact set so that

$$\{y \in \mathbb{R}^{n_2} \mid \exists x \in \mathbb{R}^{n_1} \text{ such that } (x,y) \in \tilde{G}\} \subset \tilde{G}^y \quad (7.26)$$

or even $\tilde{G}^y = \mathbb{R}^{n_2}$.

Let

$$\widehat{G}^0 := \{(x,y) \in \mathbb{R}^{n_1} \times \mathbb{R}^{n_2} \mid (x,y) \in G, \ y \in \tilde{G}^y\} \supset \widehat{G}$$

denote the constraint set of (7.25) and let

$$\widehat{\mathcal{M}}^0 := \mathcal{M}(\hat{f}(\widehat{G}^0), \hat{K})$$

be the $\hat{K}$-minimal solution set of (7.25). Then we have

$$\widehat{\mathcal{M}}^0 \cap \tilde{G} = \Omega',$$

as the following theorem shows:

Theorem 7.7. *Let $\widehat{\mathcal{M}}^0$ be the set of $\hat{K}$-minimal points of the multiobjective optimization problem (7.25) with $\hat{K} = K^1 \times \{0_{n_2}\}$ and let Ω' be the induced set of the multiobjective bilevel optimization problem (7.21) with the upper level constraint $(x,y) \in \tilde{G}$. Then $\Omega' = \widehat{\mathcal{M}}^0 \cap \tilde{G}$.*

Proof. We have the equivalences

$(\bar{x},\bar{y}) \in \Omega' \Leftrightarrow (\bar{x},\bar{y}) \in \tilde{G} \wedge (\bar{x},\bar{y}) \in G \wedge (\not\exists (x,\bar{y}) \in G \text{ with}$
$\quad f(\bar{x},\bar{y}) \in f(x,\bar{y}) + K^1 \setminus \{0_{m_1}\})$
$\Leftrightarrow (\bar{x},\bar{y}) \in \tilde{G} \wedge (\bar{x},\bar{y}) \in G \wedge \bar{y} \in \tilde{G}^y \wedge (\not\exists (x,y) \in G \text{ with}$
$\quad y = \bar{y}, \ \bar{y} \in \tilde{G}^y, \text{ and } f(\bar{x},\bar{y}) \in f(x,y) + K^1 \setminus \{0_{m_1}\})$
$\Leftrightarrow (\bar{x},\bar{y}) \in \tilde{G} \wedge (\bar{x},\bar{y}) \in \widehat{G}^0 \wedge (\not\exists (x,y) \in \widehat{G}^0,$
$\quad \text{with } \hat{f}(\bar{x},\bar{y}) \in \hat{f}(x,y) + \hat{K} \setminus \{0_{m_1+n_2}\})$
$\Leftrightarrow (\bar{x},\bar{y}) \in \tilde{G} \cap \widehat{\mathcal{M}}^0.$ □

7.6 Coupled Upper Level Constraints

Notice that we get the result of Theorem 7.7 also if we discard the constraint $y \in \tilde{G}^y$ in (7.25). However then the solution set of (7.25) may contain points $(\bar{x}, \bar{y})$ with $\bar{y} \notin \tilde{G}^y$. These points are not of interest as $\bar{y} \notin \tilde{G}^y$ implies $(\bar{x}, \bar{y}) \notin \tilde{G}$ and hence these points are not an element of $\hat{\mathcal{M}}^0 \cap \tilde{G}$. Thus, considering (7.25) without the constraint $y \in \tilde{G}^y$ makes the set $\hat{\mathcal{M}}^0$ unnecessary large.

Based on these concepts also a numerical method for solving multi-objective bilevel optimization problems with coupled upper level constraints can be developed similarly to Sect. 7.4. For more details see [70].

References

[1] M. A. Abo-Sinna. A bi-level non-linear multi-objective decision making under fuzziness. *Opsearch*, 38(5):484–495, 2001.

[2] M. Adán and V. Novo. Proper efficiency in vector optimization on real linear spaces. *J. Optimization Theory Appl.*, 121(3):515–540, 2004.

[3] M. Alber and R. Reemtsen. Intensity modulated radiotherapy treatment planning by use of a barrier-penalty multiplier method. *Optim. Methods Softw.*, 22(3):391–411, 2007.

[4] C. D. Aliprantis, M. Florenzano, V. F. Martins-da Rocha, and R. Tourky. Equilibrium analysis in financial markets with countably many securities. *J. Math. Econ.*, 40(6):683–699, 2004.

[5] C. D. Aliprantis, P. K. Monteiro, and R. Tourky. Non-marketed options, non-existence of equilibria, and nonlinear prices. *J. Econ. Theory*, 114(2):345–357, 2004.

[6] W. Alt. Parametric optimization with applications to optimal control and sequential quadratic programming. *Bayreuther Mathematische Schriften*, 35:1–37, 1991.

[7] A. Auslender and R. Cominetti. First and second order sensitivity analysis of nonlinear programs under directional constraint qualification conditions. *Optimization*, 21(3):351–363, 1990.

[8] T. Q. Bao, P. Gupta, and B. S. Mordukhovich. Necessary conditions in multiobjective optimization with equilibrium constraints. *J. Optim. Theory Appl.*, 135:179–203, 2007.

[9] J. F. Bard. *Practical bilevel optimization. Algorithms and applications.*, volume 30 of *Nonconvex Optimization and Its Applications*. Kluwer Academic Publishers, Dordrecht, 1998.

[10] H. Benson and S. Sayin. Towards finding global representations of the efficient set in multiple objective mathematical programming. *Nav. Res. Logist.*, 44(1):47–67, 1997.

[11] H. P. Benson. Existence of efficient solutions for vector maximization problems. *J. Optimization Theory Appl.*, 26:569–580, 1978.

[12] H. P. Benson. An improved definition of proper efficiency for vector maximization with respect to cones. *J. Math. Anal. Appl.*, 71:232–241, 1979.

[13] V. E. Berezkin, G. K. Kamenev, and A. V. Lotov. Hybrid adaptive methods for approximating a nonconvex multidimensional pareto frontier. *Computational Mathematics and Mathematical Physics*, 46(11):1918–1931, 2006.

[14] K. Bergstresser, A. Charnes, and P. L. Yu. Generalization of domination structures and nondominated solutions in multicriteria decision making. *J. Optimization Theory Appl.*, 18(1):3–13, 1976.

[15] H. Bernau. Interactive methods for vector optimization. In *Optimization in mathematical physics*, volume 34 of *Methoden Verfahren Math. Phys.*, pages 21–36. Pap. 11th Conf. Methods Techniques Math. Phys., Oberwolfach/Ger. 1985, 1987.

[16] E. Bijick, D. Diehl, and W. Renz. Bicriteria optimization of high frequency fields in magnetic resonance image processing. (Bikriterielle Optimierung des Hochfrequenzfeldes bei der Magnetresonanzbildgebung.). Küfer, K.-H. (ed.) et al., Multicriteria decision making and fuzzy systems. Aachen: Shaker Verlag, pages 85-98, 2006.

[17] C. Bliek, P. Spellucci, L. N. Vicente, and A. Neumaier et al. Coconut deliverable d1 algorithms for solving nonlinear constrained and optimization problems: The state of the art. http://www.mat.univie.ac.at/ neum/ms/StArt.pdf, 2001.

[18] J. F. Bonnans, R. Cominetti, and A. Shapiro. Sensitivity analysis of optimization problems under second order regular constraints. *Math. Oper. Res.*, 23(4):806–831, 1998.

[19] J. F. Bonnans and A. Shapiro. *Perturbation analysis of optimization problems*. Springer, New York, 2000.

[20] H. Bonnel and J. Morgan. Semivectorial bilevel optimization problem: Penalty approach. *J. Optim. Theory Appl.*, 131(3):365–382, 2006.

References

[21] J. M. Borwein. Proper efficient points for maximizations with respect to cones. *SIAM J. Control Optim.*, 15:57–63, 1977.

[22] J. M. Borwein. On the existence of Pareto efficient points. *Math. Oper. Res.*, 8:64–73, 1983.

[23] A. Brahme. Dosimetric precision requirements in radiation therapy. *Acta radiologica - Oncology*, 23:379–391, 1984.

[24] B. Brosowski. A criterion for efficiency and some applications. In *Optimization in mathematical physics*, volume 34 of *Methoden Verfahren Math. Phys.*, pages 37–59. Pap. 11th Conf. Methods Techniques Math. Phys., Oberwolfach/Ger. 1985, 1987.

[25] J. Buchanan and L. Gardiner. A comparison of two reference point methods in multiple objective mathematical programming. *Eur. J. Oper. Res.*, 149(1):17–34, 2003.

[26] P. H. Calamai and L. N. Vicente. Generating quadratic bilevel programming test problems. *ACM Trans. Math. Softw.*, 20(1):103–119, 1994.

[27] W. Candler and R. Norton. Multilevel programming. Technical Report 20, Technical Report of the world bank development research center, Washington, USA, 1977.

[28] V. Chankong and Y. Y. Haimes. *Multiobjective Decision Making*, volume 8 of *North Holland Series in system science and engineering*. Elsevier, Amsterdam, 1983.

[29] A. Charnes and W. Cooper. *Management models and industrial applications of linear programming*, volume 1. Wiley, New York, 1961.

[30] A. Charnes, Z. M. Huang, J. J. Rousseau, and Q. L. Wei. Cone extremal solutions of multi-payoff games with cross-constrained strategy sets. *Optimization*, 21(1):51–69, 1990.

[31] C. A. Coello Coello, G. B. Lamont, and D. A. Van Veldhuizen. *Evolutionary algorithms for solving multi-objective problems*. Genetic Algorithms and Evolutionary Computation Series. New York, NY: Springer, 2007.

[32] Y. Collette and P. Siarry. Three new metrics to measure the convergence of metaheuristics towards the pareto frontier and the aesthetic of a set of solutions in biobjective optimization. *Comput. Oper. Res.*, 32(4):773–792, 2005.

[33] B. Colson, P. Marcotte, and G. Savard. Bilevel programming: a survey. *4OR*, 3(2):87–107, 2005.

[34] B. Colson, P. Marcotte, and G. Savard. An overview of bilevel optimization. *Ann. Oper. Res.*, 153:235–256, 2007.

[35] H. W. Corley. Technical note: Games with vector payoffs. *J. Optimization Theory Appl.*, 47:491–498, 1985.

[36] C. Cotrutz, C. Lahanas, and C. Kappas. A multiobjective gradient-based dose optimization algorithm for external beam conformal radiotherapy. *Phys. Med. Biol.*, 46:2161–2175, 2001.

[37] D. Craft, T. Halabi, and T. Bortfeld. Exploration of tradeoffs in intensity-modulated radiotherapy. *Phys. Med. Biol.*, 50:5857–5868, 2005.

[38] I. Das. An improved technique for choosing parameters for pareto surface generation using normal-boundary intersection. In C. Bloebaum and K. Lewis et al., editors, *Proceedings of the Third World Congress of Structural and Multidisciplinary Optimization (WCSMO-3)*, volume 2, pages 411–413, Buffalo NY, 1999.

[39] I. Das and J. E. Dennis. A closer look at drawbacks of minimizing weighted sums of objectives for pareto set generation in multicriteria optimization problems. *Structural Optimization*, 14:63–69, 1997.

[40] I. Das and J. E. Dennis. Normal-boundary intersection: A new method for generating the pareto surface in nonlinear multicriteria optimization problems. *SIAM J. Optim.*, 8(3):631–657, 1998.

[41] K. Deb. Multi-objective genetic algorithms: Problem difficulties and construction of test problems. *Evolutionary Computation Journal*, 7(3):205–230, 1999.

[42] K. Deb. *Multi-Objective Optimization using Evolutionary Algorithms*. Wiley-Interscience Series in Systems and Optimization. John Wiley & Sons, Chichester, 2001.

[43] K. Deb and S. Jain. Running performance metrics for evolutionary multi-objective optimization. In *Proceedings of the Fourth Asia-Pacific Conference on Simulated Evolution and Learning (SEAL'02)*, pages 13–20, Singapore, 2002.

[44] K. Deb, M. Mohan, and S. Mishra. Towards a quick computation of well-spread pareto-optimal solutions. Fonseca, C.M. (ed.) et al., Evolutionary multi-criterion optimization. Second international conference, EMO 2003, Faro, Portugal, April 8–11, 2003. Proceedings. Berlin: Springer. Lect. Notes Comput. Sci. 2632, pages 222-236, 2003.

[45] K. Deb, A. Pratap, and T. Meyarivan. Constrained test problems for multi-objective evolutionary optimization. In E. Zitzler et al., editor, *First International Conference on Evolutionary Multi-Criterion Optimization*, pages 284–298, Heidelberg, 2001. Springer.
[46] K. Deb, L. Thiele, M. Laumanns, and E. Zitzler. Scalable test problems for evolutionary multiobjective optimization. In A. Abraham et al., editor, *Evolutionary Multiobjective Optimization*, pages 105–145. Springer, 2005.
[47] A. Dell'Aere. Multi-objective optimization in self-optimizing systems. In *Proceedings of the IEEE 32nd Annual Conference on Industrial Electronics (IECON), Paris*, pages 4755–4760, 2006.
[48] M. Dellnitz, O. Schütze, and T. Hestermeyer. Covering pareto sets by multilevel subdivision techniques. *J. Optim. Theory Appl.*, 124(1):113–136, 2005.
[49] S. Dempe. *Foundations of bilevel programming*, volume 61 of *Nonconvex Optimization and Its Applications*. Kluwer, Dordrecht, 2002.
[50] S. Dempe. Annotated bibliography on bilevel programming and mathematical programs with equilibrium constraints. *Optimization*, 52(3):333–359, 2003.
[51] S. Dempe. Bilevel programming - a survey. Preprint 2003-11, Fakultät für Mathematik und Informatik, TU Bergakademie Freiberg, Germany, 2003.
[52] S. Dempe, J. Dutta, and S. Lohse. Optimality conditions for bilevel programming problems. *Optimization*, 55(5-6):505–524, 2006.
[53] S. Dempe, J. Dutta, and B. S. Mordukhovich. New necessary optimality conditions in optimistic bilevel programming. *Optimization*, 56(5-6):577–604, 2007.
[54] S. Deng. On efficient solutions in vector optimization. *J. Optimization Theory Appl.*, 96(1):201–209, 1998.
[55] W. Dinkelbach. *Sensitivitätsanalysen und parametrische Programmierung*. Ökonometrie und Unternehmensforschung. 12. Berlin-Heidelberg-New York: Springer-Verlag. XI, 1969.
[56] W. Dinkelbach and W. Dürr. Effizienzaussagen bei Ersatzprogrammen zum Vektormaximumproblem. In *Operations Res.-Verf. 12, IV. Oberwolfach-Tag. Operations Res. 1971*, pages 69–77. 1972.

[57] K. Doerner, W. J. Gutjahr, R. F. Hartl, C. Strauss, and C. Stummer. Pareto ant colony optimization: A metaheuristic approach to multiobjective portfolio selection. *Ann. Oper. Res.*, 131:79–99, 2004.

[58] G. Dorini, F. Di Pierro, D. Savic, and A. B. Piunovskiy. Neighbourhood search for constructing pareto sets. *Math. Meth. Oper. Res.*, 65(2):315–337, 2007.

[59] F. Y. Edgeworth. *Mathematical Psychics*. University Microfilms International (Out-of-Print Books on Demand), 1987 (the original edition in 1881).

[60] M. Ehrgott. *Multicriteria optimisation*, volume 491 of *Lect. Notes Econ. math. Syst.* Springer, Berlin, 2000.

[61] M. Ehrgott. An optimisation model for intensity modulated radiation therapy. In *Proceedings of the 37th Annual Conference of the Operational Research Society of New Zealand*, pages 23–31, Auckland, 2002.

[62] M. Ehrgott. *Multicriteria optimisation*. Springer, Berlin, 2nd edition, 2005.

[63] M. Ehrgott and M. Burjony. Radiation therapy planning by multicriteria optimization. In *Proceedings of the 36th Annual Conference of the Operational Research Society of New Zealand*, pages 244–253, Auckland, 2001.

[64] M. Ehrgott and R. Johnston. Optimisation of beam directions in intensity modulated radiation therapy planning. *OR Spectrum*, 25(2):251–264, 2003.

[65] M. Ehrgott, K. Klamroth, and C. Schwehm. An mcdm approach to portfolio optimization. *Eur. J. Oper. Res.*, 155(3):752–770, 2004.

[66] M. Ehrgott and M. M. Wiecek. Multiobjective programming. Figueira, J. (ed.) et al., Multiple criteria decision analysis. State of the art surveys. New York, NY: Springer, pages 667-722, 2005.

[67] G. Eichfelder. ε-constraint method with adaptive parameter control and an application to intensity-modulated radiotherapy. Küfer, K.-H. (ed.) et al., Multicriteria decision making and fuzzy systems. Aachen: Shaker Verlag, pages 25-42, 2006.

[68] G. Eichfelder. Multiobjective bilevel optimization. Preprint-Series of the Institute of Applied Mathematics 312, Institute of Applied Mathematics, Univ. Erlangen-Nürnberg, Germany, 2007.

[69] G. Eichfelder. Scalarizations for adaptively solving multi-objective optimization problems. *Comput. Optim. Appl.*, 2007. DOI: 10.1007/s10589-007-9155-4.

[70] G. Eichfelder. Solving nonlinear multiobjective bilevel optimization problems with coupled upper level constraints. Preprint-Series of the Institute of Applied Mathematics 320, Institute of Applied Mathematics, Univ. Erlangen-Nürnberg, Germany, 2007.

[71] A. Engau. Variable preference modeling with ideal-symmetric convex cones. *J. Glob. Optim.*, 2007. DOI 10.1007/s10898-007-9246-x.

[72] A. Engau and M. M. Wiececk. Introducing non-polyhedral cones to multi-objective programming. In *Book of Abstracts of the 7th Int. Conf. on Multi-Objective Programming and Goal Programming*, 2006.

[73] A. Engau and M. M. Wiecek. Cone characterizations of approximate solutions in real vector optimization. *J. Optim. Theory Appl.*, 134:499–513, 2007.

[74] A. Engau and M. M. Wiecek. Generating ε-efficient solutions in multiobjective programming. *Eur. J. Oper. Res.*, 177(3):1566–1579, 2007.

[75] J. Ester. *Systemanalyse und mehrkriterielle Entscheidung*. Verlag Technik, Berlin, 1987.

[76] G. W. Evans. An overview of techniques for solving multiobjective mathematical programs. *Manage. Sci.*, 30:1268–1282, 1984.

[77] J. Fernández and B. Tóth. Obtaining an outer approximation of the efficient set of nonlinear biobjective problems. *J. Global Optim.*, 38(2):315–331, 2007.

[78] A. V. Fiacco. Sensitivity analysis for nonlinear programming using penalty methods. *Math. Program.*, 10:287–311, 1976.

[79] A. V. Fiacco. *Introduction to Sensitivity and Stability Analysis in Nonlinear Programming*, volume 165 of *Mathematics in Science and Engineering*. Academic Press, London, 1983.

[80] R. Fletcher. *Practical methods of optimization, 2nd ed.* Wiley, 1987.

[81] J. Fliege. Gap-free computation of pareto-points by quadratic scalarizations. *Math. Methods Oper. Res.*, 59(1):69–89, 2004.

[82] J. Fliege. An efficient interior-point method for convex multicriteria optimization problems. *Math. Oper. Res.*, 31(4):825–845, 2006.

[83] J. Fliege, C. Heermann, and B. Weyers. A new adaptive algorithm for convex quadratic multicriteria optimization. In J. Branke, K. Deb, K. Miettinen, and R. E. Steuer, editors, *Practical Approaches to Multi-Objective Optimization*, number 04461 in Dagstuhl Seminar Proceedings. Schloss Dagstuhl, Germany, 2005.

[84] J. Fliege and A. Heseler. Constructing approximations to the efficient set of convex quadratic multiobjective problems. Ergebnisberichte Angewandte Mathematik 211, Angewandte Mathematik, Univ. Dortmund, Germany, 2002.

[85] J. Fliege and L. N. Vicente. Multicriteria approach to bilevel optimization. *J. Optimization Theory Appl.*, 131(2):209–225, 2006.

[86] J. Focke. Vektormaximumproblem und parametrische Optimierung. *Math. Operationsforsch. Stat.*, 4:365–369, 1973.

[87] C. M. Fonseca and P. J. Fleming. An overview of evolutionary algorithms in multiobjective optimization. *Evolutionary Computation*, 3(1):1–16, 1995.

[88] J. Fortuny-Amat and B. McCarl. A representation and economic interpretation of a two-level programming problem. *J. Oper. Res. Soc.*, 32:783–792, 1981.

[89] F. W. Gembicki and Y. Y. Haimes. Approach to performance and sensitivity multiobjective optimization: The goal attainment method. *IEEE Trans. Automatic Control*, 6:769–771, 1975.

[90] A. M. Geoffrion. Proper efficiency and the theory of vector maximization. *J. Math. Anal. Appl.*, 22:618–630, 1968.

[91] C. Gerstewitz. Nichtkonvexe Dualität in der Vektoroptimierung. *Wissensch. Zeitschr. TH Leuna-Merseburg*, 25(3):357–364, 1983.

[92] C. Gerth and P. Weidner. Nonconvex separation theorems and some applications in vector optimization. *J. Optimization Theory Appl.*, 67(2):297–320, 1990.

[93] C. Gil, A. Márquez, R. Baños, M. G. Montoya, and J. Gómez. A hybrid method for solving multi-objective global optimization problems. *J. Global Optim*, 38(2):265–281, 2007.

[94] A. Goepfert and R. Nehse. *Vektoroptimierung: Theorie, Verfahren und Anwendungen*. Teubner, Leipzig, 1990.

[95] D. Gourion and D. T. Luc. Generating the weakly efficient set of nonconvex multiobjective problems. *J. Global Optim.*, 2007. DOI 10.1007/s10898-007-9263-9.

[96] L. M. Graña Drummond and B. F. Svaiter. A steepest descent method for vector optimization. *J. Comput. Appl. Math.*, 175(2):395–414, 2005.

[97] J. Guddat, F. Guerra Vasquez, K. Tammer, and K. Wendler. *Multiobjective and stochastic optimization based on parametric optimization.* Akademie-Verlag, Berlin, 1985.

[98] Y. Y. Haimes, L. S. Lasdon, and D. A. Wismer. On a bicriterion formulation of the problems of integrated system identification and system optimization. *IEEE Trans. Syst. Man Cybern.*, 1:296–297, 1971.

[99] W. A. Hall and Y. Y. Haimes. The surrogate worth trade-off method with multiple decision-makers. In M. Zeleny, editor, *Multiple Crit. Decis. Making, 22nd int. Meet. TIMS, Kyoto 1975*, volume 123 of *Lect. Notes Econ. math. Syst.*, pages 207–233. Springer, Berlin, 1976.

[100] H. W. Hamacher and K.-H. Küfer. Inverse radiation therapy planning - a multiple objective optimization approach. *Discrete Appl. Math.*, 118:145–161, 2002.

[101] M. P. Hansen and A. Jaszkiewicz. Evaluating the quality of approximations to the non-dominated set. Technical Report IMM-REP-1998-7, IMM Technical Univ. of Denmark, 1998.

[102] R. Hartley. On cone-efficiency, cone-convexity and cone-compactness. *SIAM Journal on Applied Mathematics*, 34(2):211–222, 1978.

[103] G. B. Hazen. Differential characterizations of nonconical dominance in multiple objective decision making. *Math. Oper. Res.*, 13(1):174–189, 1988.

[104] S. Helbig. An interactive algorithm for nonlinear vector optimization. *Appl. Math. Optimization*, 22(2):147–151, 1990.

[105] S. Helbig. On the connectedness of the set of weakly efficient points of a vector optimization problem in locally convex spaces. *J. Optimization Theory Appl.*, 65(2):257–270, 1990.

[106] S. Helbig. Approximation of the efficient point set by perturbation of the ordering cone. *Z. Oper. Res.*, 35(3):197–220, 1991.

[107] M. I. Henig. Proper efficiency with respect to cones. *J. Optimization Theory Appl.*, 36:387–407, 1982.

[108] M. I. Henig. Value functions, domination cones and proper efficiency in multicriteria optimization. *Math. Program., Ser. A*, 46(2):205–217, 1990.

[109] A. Heseler. *Warmstart-Strategien in der parametrisierten Optimierung mit Anwendung in der multikriteriellen Optimierung.* dissertation, Univ. Dortmund, Germany, 2005.

[110] C. Hillermeier. Generalized homotopy approach to multiobjective optimization. *J. Optimization Theory Appl.*, 110(3):557–583, 2001.

[111] C. Hillermeier. *Nonlinear multiobjective optimization. A generalized homotopy approach.* Birkhäuser, 2001.

[112] C. Hillermeier and J. Jahn. Multiobjective optimization: survey of methods and industrial applications. *Surv. Math. Ind.*, 11:1–42, 2005.

[113] M. Hirschberger. *Struktur effizienter Punkte in Quadratischen Vektoroptimierungsproblemen.* Berichte aus der Mathematik. Shaker, Aachen, 2003.

[114] A. L. Hoffmann, A. Y. D. Siem, D. den Hertog, J. Kaanders, and H. Huizenga. Derivative-free generation and interpolation of convex pareto optimal imrt plans. *Phys. Med. Biol.*, 51:6349–6369, 2006.

[115] R. Horst and H. Tuy. *Global Optimization: Deterministic Approaches.* Springer, Berlin, 1990.

[116] B. J. Hunt. *Multiobjective Programming with Convex Cones: Methodology and Applications.* Phd thesis, Univ. of Clemson, USA, 2004.

[117] B. J. Hunt and M. M. Wiecek. Cones to aid decision making in multicriteria programming. In T. Tanino, T. Tanaka, and M. Inuiguchi, editors, *Multi-Objective Programming and Goal-Programming*, pages 153–158. Springer, Berlin, 2003.

[118] A. Hutterer and J. Jahn. On the location of antennas for treatment planning in hyperthermia. *OR Spectrum*, 25(3):397–412, 2003.

[119] C.-L. Hwang and A. S. M. Masud. *Multiple objective decision making - methods and applications. A state-of-the-art survey*, volume 164 of *Lect. Notes Econ. math. Syst.* Springer-Verlag, Berlin, 1979.

[120] J. Jahn. Scalarization in vector optimization. *Math. Program.*, 29:203–218, 1984.

[121] J. Jahn. A characterization of properly minimal elements of a set. *SIAM J. Control Optimization*, 23:649–656, 1985.

References 229

[122] J. Jahn. *Mathematical Vector Optimization in Partially Ordered Linear Spaces*. Lang, Frankfurt, 1986.
[123] J. Jahn. *Introduction to the theory of nonlinear optimization*. Springer, Berlin, 1994.
[124] J. Jahn. *Vector Optimization: Theory, Applications and Extensions*. Springer, Berlin, 2004.
[125] J. Jahn. Multiobjective search algorithm with subdivision technique. *Comput. Optim. Appl.*, 35:161–175, 2006.
[126] J. Jahn, A. Kirsch, and C. Wagner. Optimization of rod antennas of mobile phones. *Math. Methods Oper. Res.*, 59(1):37–51, 2004.
[127] J. Jahn, J. Klose, and A. Merkel. On the application of a method of reference point approximation to bicriterial optimization problems in chemical engineering. Advances in optimization, Proc. 6th Fr.-Ger. Colloq., Lambrecht/Ger. 1991, Lect. Notes Econ. Math. Syst. 382, pages 478-491, 1992.
[128] J. Jahn and A. Merkel. Reference point approximation method for the solution of bicriterial nonlinear optimization problems. *J. Optimization Theory Appl.*, 74(1):87–103, 1992.
[129] J. Jahn and E. Schaller. A global solver for multiobjective nonlinear bilevel optimization problems. Preprint-Series of the Institute of Applied Mathematics 318, Univ. Erlangen-Nürnberg, Germany, 2007.
[130] R. Janka. Private communication, 2004.
[131] K. Jittorntrum. Solution point differentiability without strict complementarity in nonlinear programming. *Math. Program. Study*, 21:127–138, 1984.
[132] D. R. Jones, C. D. Perttunen, and B. E. Stuckman. Lipschitzian optimization without the lipschitz constant. *J. Optimization Theory Appl.*, 79(1):157–181, 1993.
[133] A. Jüschke, J. Jahn, and A. Kirsch. A bicriterial optimization problem of antenna design. *Comput. Optim. Appl.*, 7(3):261–276, 1997.
[134] I. Kaliszewski. *Quantitative pareto analysis by cone separation technique*. Kluwer Academic Publishers, Boston, 1994.
[135] I. Kaliszewski. Trade-offs – a lost dimension in multiple criteria decision making. Trzaskalik, T. (ed.) et al., Multiple objective and goal programming. Recent developments. Proceedings of the 4th international conference, (MOPGP'00), Ustroń, Poland, May 29-June 1, 2000. Heidelberg: Physica-Verlag, pages 115-126, 2002.

230 References

[136] R. Kasimbeyli. An approximation of convex cones in banach spaces and scalarization in nonconvex vector optimization. manuscript, 2007.

[137] G. A. Kiker, T. S. Bridges, A. Varghese, T. P. Seager, and I. Linkov. Application of multicriteria decision analysis in environmental decision making. *Integrated Environmental Assessment and Management*, 1(2):95–108, 2005.

[138] I. Y. Kim and O. de Weck. Adaptive weighted sum method for bi-objective optimization. *Structural and Multidisciplinary Optimization*, 29:149 – 158, 2005.

[139] I. Y. Kim and O. de Weck. Adaptive weighted sum method for multiobjective optimization: a new method for pareto front generation. *Structural and Multidisciplinary Optimization*, 31:105–116, 2006.

[140] J. Kim and S.-K. Kim. A chim-based interactive tchebycheff procedure for multiple objective decision making. *Comput. Oper. Res.*, 33(6):1557–1574, 2006.

[141] J. Knowles and D. Corne. On metrics for comparing nondominated sets. In *Proceedings of the World Congress on Computational Intelligence*, pages 711–716, 2002.

[142] K.-H. Küfer and M. Monz. Private communication, 2004.

[143] K.-H. Küfer, A. Scherrer, M. Monz, F. Alonso, H. Trinkaus, T. Bortfeld, and C. Thieke. Intensity-modulated radiotherapy - a large scale multi-criteria programming problem. *OR Spectrum*, 25:223–249, 2003.

[144] H. W. Kuhn and A. W. Tucker. Nonlinear programming. In J. Neyman, editor, *Proceedings of the second Berkeley Symposium on mathematical Statistics and Probability*, pages 481–492, Berkeley, 1951. Univ. of California Press.

[145] M. Labbé, P. Marcotte, and G. Savard. On a class of bilevel programs. In G. Di Pillo, editor, *Nonlinear optimization and related topics. Workshop, Erice, Sicily, Italy, June 23–July 2, 1998*, volume 36 of *Appl. Optim.*, pages 183–206, Dordrecht, 2000. Kluwer Academic Publishers.

[146] M. Lahanas. History of multi-criteria (or multiobjective) optimization in radiation therapy. http://www.mlahanas.de/MedPhys/MORadiotherapy.htm, 2005.

[147] M. Laumanns, L. Thiele, and E. Zitzler. An adaptive scheme to generate the pareto front based on the epsilon-constraint method. In J. Branke, K. Deb, K. Miettinen, and R. E. Steuer, editors, *Practical Approaches to Multi-Objective Optimization*, number 04461 in Dagstuhl Seminar Proceedings. Schloss Dagstuhl, Germany, 2005.

[148] M. Laumanns, L. Thiele, and E. Zitzler. An efficient, adaptive parameter variation scheme for metaheuristics based on the epsilon-constraint method. *European J. Oper. Res.*, 169(3):932–942, 2006.

[149] D. Li, J.-B. Yang, and M. P. Biswal. Quantitative parametric connections between methods for generating noninferior solutions in multiobjective optimization. *EJOR*, 117:84–99, 1999.

[150] J. G. Lin. Maximal vectors and multi-objective optimization. *J. Optimization Theory Appl.*, 18:41–64, 1976.

[151] J. G. Lin. On min-norm and min-max methods of multi-objective optimization. *Math. Program.*, 103(1):1–33, 2005.

[152] S. A. Y. Lin. A comparison of Pareto optimality and domination structure. *Metroeconomica*, 28:62–74, 1976.

[153] Y. C. Liou, X. Q. Yang, and J. C. Yao. Mathematical programs with vector optimization constraints. *J. Optim. Theory Appl.*, 126(2):345–355, 2005.

[154] P. Loridan. ε-solutions in vector minimization problems. *J. Optimization Theory Appl.*, 43:265–276, 1984.

[155] K. Löwner. Über monotone Matrixfunktionen. *Math. Z.*, 38:177–216, 1934.

[156] D. T. Luc. Connectedness of the efficient point sets in quasiconcave vector maximization. *J. Math. Anal. Appl.*, 122:346–354, 1987.

[157] D. G. Luenberger. *Introduction to linear and nonlinear programming*. Addison-Wesley, Reading, Mass., 1973.

[158] M. Luque, K. Miettinen, P. Eskelinen, and F. Ruiz. Three different ways for incorporating preference information in interactive reference point based methods. working paper W-410, Helsinki School of Economics, Finland, 2006.

[159] S. A. Marglin. *Public investment criteria*. MIT Press, Cambridge, 1967.

[160] C. A. Mattson and A. Messac. Concept selection using s-pareto frontiers. *AIAA Journal*, 41(6):1190–1198, 2003.

[161] C. A. Mattson, A. A. Mullur, and A. Messac. Smart pareto filter: Obtaining a minimal representation of multiobjective design space. *Engineering Optimization*, 36(6):721–740, 2004.

[162] H. Maurer and J. Zowe. First and second-order necessary and sufficient optimality conditions for infinite-dimensional programming problems. *Math. Program.*, 16:98–110, 1979.

[163] A. Messac, A. Ismail-Yahaya, and C. A. Mattson. The normalized normal constraint method for generating the pareto frontier. *Struct. Multidisc.Optim.*, 25(2):86–98, 2003.

[164] A. Messac and C. A. Mattson. Normal constraint method with guarantee of even representation of complete pareto frontier. *AIAA Journal*, 42(10):2101–2111, 2004.

[165] K. Miettinen. *Nonlinear multiobjective optimization*. Kluwer Academic Publishers, Boston, 1999.

[166] K. Miettinen. Some methods for nonlinear multi-objective optimization. In E. Zitzler et al., editor, *Evolutionary Multi-Criterion Optimization: First International Conference, EMO 2001, Zurich, Switzerland, March 2001. Proceedings*, volume 1993 of *Lecture Notes in Computer Science*, pages 1–20, 2001.

[167] K. Miettinen and M. M. Mäkelä. On scalarizing functions in multiobjective optimization. *OR Spectrum*, 24(2):193–213, 2002.

[168] A. Migdalas. Bilevel programming in traffic planning: Models, methods and challenge. *J. Glob. Optim.*, 7(4):381–405, 1995.

[169] E. Miglierina and E. Molho. Scalarization and stability in vector optimization. *J. Optimization Theory Appl.*, 114(3):657–670, 2002.

[170] M. Monz. *Pareto navigation: interactive multiobjective optimisation and its application in radiotherapy planning.* dissertation, Technische Univ. Kaiserslautern, Germany, 2006.

[171] B. S. Mordukhovich. Multiobjective optimization problems with equilibrium constraints. *Math. Programming*, 2007. DOI 10.1007/s10107-007-0172-y.

[172] P. H. Naccache. Connectedness of the set of nondominated outcomes im multicriteria optimization. *J. Optimization Theory Appl.*, 25:459–467, 1978.

[173] H. Nakayama. Trade-off analysis based upon parametric optimization. Multiple criteria decision support, Proc. Int. Workshop, Helsinki/Finl. 1989, Lect. Notes Econ. Math. Syst. 356, pages 42-52, 1991.

[174] H. Nakayama. Some remarks on trade-off analysis in multi-objective programming. Clímaco, J. (ed.), Multicriteria analysis. Proceedings of the XIth international conference on MCDM, Coimbra, Portugal, August 1–6, 1994. Berlin: Springer. pages 179-190, 1997.

[175] H. Nakayama. Multi-objective optimization and its engineering applications. In J. Branke, K. Deb, K. Miettinen, and R. E. Steuer, editors, *Practical Approaches to Multi-Objective Optimization*, number 04461 in Dagstuhl Seminar Proceedings. Schloss Dagstuhl, Germany, 2005.

[176] A. Niemierko. Reposting and analysing dose distributions: a concept of equivalent uniform dose. *Medical Physics*, 24:103–110, 1997.

[177] I. Nishizaki and T. Notsu. Nondominated equilibrium solutions of a multiobjective two-person nonzero-sum game and corresponding mathematical programming problem. *J. Optim. Theory Appl.*, 135(2):217–269, 2007.

[178] I. Nishizaki and M. Sakawa. Stackelberg solutions to multiobjective two-level linear programming problems. *J. Optimization Theory Appl.*, 103(1):161–182, 1999.

[179] M. S. Osman, M. A. Abo-Sinna, A. H. Amer, and O. E. Emam. A multi-level nonlinear multi-objective decision-making under fuzziness. *Appl. Math. Comput.*, 153(1):239–252, 2004.

[180] V. Pareto. *Cours d'Economie Politique*. Libraire Droz, Genève, 1964 (the first edition in 1896).

[181] A. Pascoletti and P. Serafini. Scalarizing vector optimization problems. *J. Optimization Theory Appl.*, 42(4):499–524, 1984.

[182] E. Polak. On the approximation of solutions to multiple criteria decision making problems. In M. Zeleny, editor, *Multiple Crit. Decis. Making, 22nd int. Meet. TIMS, Kyoto 1975*, volume 123 of *Lect. Notes Econ. math. Syst.*, pages 271–282. Springer, Berlin, 1976.

[183] M. J. D. Powell. *The Convergence of Variable Metric Methods for Nonlinearly Constrained Optimization Calculations*, volume 3 of *Nonlinear Programming*. Academic Press, 1978.

[184] J. Prohaska. Optimierung von Spulenkonfigurationen zur Bewegung magnetischer Sonden. Diplomarbeit, Univ. Erlangen-Nürnberg, Germany, 2005.

[185] R. Ramesh, M. H. Karwan, and S. Zionts. Theory of convex cones in multicriteria decision making. Multi-attribute decision making via O.R.-based expert systems, Proc. Int. Conf., Passau/FRG 1986, Ann. Oper. Res. 16, No.1-4, pages 131-147, 1988.

[186] S. R. Ranjithan, S. K. Chetan, and H. K. Dakshina. Constraint method-based evolutionary algorithm (cmea) for multiobjective optimization. In E. Zitzler et al., editor, *Evolutionary Multi-Criterion Optimization: First International Conference, EMO 2001, Zurich, Switzerland, March 2001. Proceedings*, volume 1993 of *Lecture Notes in Computer Science*, pages 299–313, 2001.

[187] S. M. Robinson. Strongly regular generalized equations. *Math. Oper. Res.*, 5(1):43–62, 1980.

[188] S. Ruzika. Approximation algorithms in multiple objective programming. Master's thesis, Univ. Kaiserslautern, Germany, 2003.

[189] S. Ruzika and M. M. Wiecek. Approximation methods in multiobjective programming. *J. Optimization Theory Appl.*, 126(3):473–501, 2005.

[190] Y. Sawaragi, H. Nakayama, and T. Tanino. *Theory of Multiobjective Optimization*. Number 176 in Mathematics in science and engineering. Academic Press, London, 1985.

[191] S. Sayin. Measuring the quality of discrete representations of efficient sets in multiple objective mathematical programming. *Math. Program.*, 87,A:543–560, 2000.

[192] S. Schäffler. *Global Optimization Using Stochastic Integration*. S.Roderer Verlag, Regensburg, 1995.

[193] S. Schäffler and K. Ritter. A stochastic method for constrained global optimization. *SIAM J. Optim.*, 4(4):894–904, 1994.

[194] S. Schäffler, R. Schultz, and K. Weinzierl. Stochastic method for the solution of unconstrained vector optimization problems. *J. Optimization Theory Appl.*, 114(1):209–222, 2002.

[195] E. Schaller. Ein Verfahren zur globalen Lösung multikriterieller nichtlinearer Zwei-Ebenen-Optimierungsprobleme. Diplomarbeit, Univ. Erlangen-Nürnberg, Germany, 2006.

[196] B. Schandl, K. Klamroth, and M. M. Wiecek. Norm-based approximation in bicriteria programming. *Comput. Optim. Appl.*, 20(1):23–42, 2001.

[197] B. Schandl, K. Klamroth, and M. M. Wiecek. Introducing oblique norms into multiple criteria programming. *J. Glob. Optim.*, 23(1):81–97, 2002.

[198] A. Scherrer, K.-H. Küfer, T. Bortfeld, M. Monz, and F. Alonso. Imrt planning on adaptive volume structures—a decisive reduction in computational complexity. *Phys. Med. Biol.*, 50:2033–2053, 2005.

[199] K. Schittkowski. Nlqpl: A fortran-subroutine solving constrained nonlinear programming problems. *Annals of Operations Research*, 5:485–500, 1985.

[200] E. Schreibmann, M. Lahanas, and L. Xing. Multiobjective evolutionary optimization of the number of beams, their orientations and weights for intensity-modulated radiation therapy. *Phys. Med. Biol.*, 49:747–770, 2004.

[201] O. Schütze, A. Dell'Aere, and M. Dellnitz. On continuation methods for the numerical treatment of multi-objective optimization problems. In J. Branke, K. Deb, K. Miettinen, and R. E. Steuer, editors, *Practical Approaches to Multi-Objective Optimization*, number 04461 in Dagstuhl Seminar Proceedings. Schloss Dagstuhl, Germany, 2005.

[202] A. Seeger. Second order directional derivatives in parametric optimization problems. *Math. Oper. Res.*, 13(1):124–139, 1988.

[203] A. Shapiro. Second-order derivatives of extremal-value functions and optimality conditions for semi-infinite programs. *Math. Oper. Res.*, 10:207–219, 1985.

[204] A. Shapiro. Second order sensitivity analysis and asymptotic theory of parametrized nonlinear programs. *Math. Program.*, 33:280–299, 1985.

[205] L. S. Shapley. Equilibrium points in games with vector payoffs. *Nav. Res. Logist. Q.*, 6:57–61, 1959.

[206] X. Shi and H. Xia. Interactive bilevel multi-objective decision making. *J. Oper. Res. Soc.*, 48(9):943–949, 1997.

[207] X. Shi and H. Xia. Model and interactive algorithm of bi-level multi-objective decision-making with multiple interconnected decision makers. *J. Multi-Criteria Decis. Anal.*, 10:27–34, 2001.

[208] P. K. Shukla. On the normal boundary intersection method for generation of efficient front. In Y. Shi, G. D. van Albada, J. Dongarra, and P. M. A. Sloot, editors, *Computational Science - ICCS 2007, 7th International Conference Beijing, China*, volume 4487 of *Lecture Notes in Computer Science*, pages 310–317. Springer, 2007.

[209] P. K. Shukla and K. Deb. On finding multiple pareto-optimal solutions using classical and evolutionary generating methods. *Eur. J. Oper. Res.*, 181(3):1630–1652, 2007.

[210] A. Y. D. Siem, D. den Hertog, and A. L. Hoffmann. The effect of transformations on the approximation of univariate (convex) functions with applications to pareto curves. Technical report, CentER Discussion Paper Series No. 2006-66, Tilburg Univ., 2006.

[211] W. Stadler. *Multicriteria optimization in engineering and in the sciences*, volume 37 of *Mathematical Concepts and Methods in Science and Engineering*. Plenum Press, New York, 1988.

[212] T. Staib. On two generalizations of pareto minimality. *J. Optimization Theory Appl.*, 59(2):289–306, 1988.

[213] A. Sterna-Karwat. Continuous dependence of solutions on a parameter in a scalarization method. *J. Optimization Theory Appl.*, 55(3):417, 434 1987.

[214] A. Sterna-Karwat. Lipschitz and differentiable dependence of solutions on a parameter in a scalarization method. *J. Aust. Math. Soc., Ser. A*, 42:353–364, 1987.

[215] R. E. Steuer. *Multiple criteria optimization. Theory, computation, and application*. Wiley, 1986.

[216] R. E. Steuer and E.-U. Choo. An interactive weighted tchebycheff procedure for multiple objective programming. *Math. Program.*, 26:326–344, 1983.

[217] R. E. Steuer and P. Na. Multiple criteria decision making combined with finance: a categorized bibliographic study. *Eur. J. Oper. Res.*, 150(3):496–515, 2003.

[218] D. Sun and J. Sun. Semismooth matrix-valued functions. *Math. Oper. Res.*, 27(1):150–169, 2002.

[219] C. Tammer and K. Winkler. A new scalarization approach and applications in multicriteria d. c. optimization. *J. Nonlinear Convex Anal.*, 4(3):365–380, 2003.

[220] K. Tamura and S. Miura. Necessary and sufficient conditions for local and global nondominated solutions in decision problems with multi-objectives. *J. Optimization Theory Appl.*, 28:501–521, 1979.

[221] M. Tanaka. Ga-based decision support system for multi-criteria optimization. In *Proceedings of the international Conference on Systems, Man and Cybernetics 2*, pages 1556–1561, 1995.

[222] C.-X. Teng, L. Li, and H.-B. Li. A class of genetic algorithms on bilevel multi-objective decision making problem. *J. of Systems Science and Systems Engineering*, 9(3):290–296, 2000.

[223] C. Thieke. *Multicriteria Optimization in Inverse Radiotherapy Planning*. dissertation, Univ. Heidelberg, Germany, 2003.

[224] D. F. Thompson, S. Gupta, and A. Shukla. Tradeoff analysis in minimum volume design of multi-stage spur gear reduction units. *Mech. Mach. Theory*, 35(5):609–627, 2000.

[225] A. Törn and A. Žilinskas. *Global Optimization*, volume 350 of *Lect. Notes in Computer Science*. Springer, Berlin, 1989.

[226] H. Tuy, A. Migdalas, and N. T. Hoai-Phuong. A novel approach to bilevel nonlinear programming. *J. Global Optim.*, 38(4):527–554, 2007.

[227] K. Vaillancourt and J.-P. Waaub. Environmental site evaluation of waste management facilities embedded into eugène model: A multicriteria approach. *Eur. J. Oper. Res.*, 139(2):436–448, 2002.

[228] D. A. Van Veldhuizen. *Multiobjective Evolutionary Algorithms: Classifications, Analyses, and New Innovation*. PhD thesis, Graduate School of Engineering, Air Force Institute of Technology, Dayton, Ohio, 1999.

[229] L. N. Vicente and P. H. Calamai. Bilevel and multilevel programming: A bibliography review. *J. Glob. Optim.*, 5(3):291–306, 1994.

[230] P. Weidner. Dominanzmengen und Optimalitätsbegriffe in der Vektoroptimierung. *Wiss. Z. Tech. Hochsch. Ilmenau*, 31(2):133–146, 1985.

[231] P. Weidner. *Ein Trennungskonzept und seine Anwendung auf Vektoroptimierungsprobleme*. Professorial dissertation, Univ. of Halle, Germany, 1990.

[232] P. Weidner. Tradeoff directions and dominance sets. Tanino, T. (ed.) et al., Multi-objective programming and goal-programming. Theory and applications. Proceeedings of the conference MOPGP'02, Nara, Japan, June 4–7, 2002. Berlin: Springer, pages 275-280, 2003.

[233] R. E. Wendell and D. N. Lee. Efficiency in multiple objective optimization problems. *Math. Program.*, 12:406–414, 1977.

[234] D. J. White. Epsilon-dominating solutions in mean-variance portfolio analysis. *Eur. J. Oper. Res.*, 105(3):457–466, 1998.

[235] M. M. Wiecek. Multi-scenario multi-objective optimization with applications in engineering design. Semi-plenary talk given at the 7th International Conference devoted to Multi-Objective Programming and Goal Programming, Tours, France (2006).

[236] A. P. Wierzbicki. Reference point approaches. Gal, T. (ed.) et al., Multicriteria decision making: Advances in MCDM models, algorithms, theory, and applications. Boston: Kluwer Academic Publishers. Int. Ser. Oper. Res. Manag. Sci. 21, 9.1-9.39, 1999.

[237] K. Winkler. Skalarisierung mehrkriterieller Optimierungsprobleme mittels schiefer Normen. Habenicht, W. (ed.) et al., Multicriteria and fuzzy systems in theory and practice. Wiesbaden: DUV, Deutscher Universitäts-Verlag, pages 173-190, 2003.

[238] H. C. Wu. A solution concept for fuzzy multiobjective programming problems based on convex cones. *J. Optimization Theory Appl.*, 121(2):397–417, 2004.

[239] S. W. Xiang and W. S. Yin. Stability results for efficient solutions of vector optimization problems. *J. Optimization Theory Appl.*, 134:385–398, 2007.

[240] J. J. Ye and Q. J. Zhu. Multiobjective optimization problem with variational inequality constraints. *Math. Programming*, 96(1):139–160, 2003.

[241] Y. Yin. Multiobjective bilevel optimization for transportation planning and management problems. *J. Advanced Transportation*, 36(1):93–105, 2000.

[242] Y. M. Younes. *Studies on discrete vector optimization.* dissertation, Univ. of Demiatta, Egypt, 1993.

[243] P. L. Yu. Cone convexity, cone extreme points, and nondominated solutions in decision problems with multiple objectives. *J. Optimization Theory Appl.*, 14:319–377, 1974.

[244] P.-L. Yu. *Multiple-criteria decision making. Concepts, techniques, and extensions. With the assistance of Yoon-Ro Lee and Antonie Stam.* Mathematical Concepts and Methods in Science and Engineering, 30. New York-London: Plenum Press, 1985.

[245] L. Zadeh. Optimality and non-scalared-valued performance criteria. *IEEE Trans. Automatic Control*, 8:59–60, 1963.

[246] E. Zitzler. *Evolutionary Algorithms for Multiobjective Optimization: Methods and Application.* PhD thesis, Swiss Federal Institute of Technology (ETH), Zürich, Switzerland, 1999.

Index

adaptive ε-constraint method, 129, 138, 142, 146, 175
adaptive modified Polak method, 133, 147
adaptive NBI method, 132, 146
adaptive Pascoletti-Serafini method, 124
algebraic interior, 11
   relative, 11, 28
antisymmetric, 5

beamlet, 168
Benson's method, 66
bilevel optimization, 183
   multiobjective, 183
   semivectorial, 183
binary relation, 5
boundary, 14, 26

cardinality, 105
CHIM, 54
cluster, 169
cone, 3
   convex, 3
   dual, 21, 189
   finitely generated, 9, 15
   ice cream, 9, 19
   ordering, 5
   pointed, 4
   polyhedral, 9, 15, 154
connected set, 104, 149
convex hull of individual minima (CHIM), 54
core, 11
   intrinsic, 11, 28, 193
coverage error, 103
   modified, 104

decision maker (d. m.), 4, 134, 185
degenerated, 85
derivative
   directional, 92
   Fréchet-, 69
dominate, 6
domination structure, 4

efficient point, 6
   properly, 12, 65
   weakly, 10
efficient set, 6
   weakly, 10
elastic constraint method, 66
ε-constraint problem, 21, 49, 94, 128, 141, 175
equilibrium problem, 184

equivalent uniform dose (EUD), 170
evolutionary algorithms, 22, 103
extremal point, 54

follower, 184

goal, 58
goal attainment method, 58
Gourion-Luc problem, 58
Graef-Younes method, 193

homogeneity, 176
hybrid method, 66
hyperbola efficiency problem, 66

ideal point, 53
IMRT, 167
induced set, 187
intensity modulated radiotherapy (IMRT), 167
interactive methods, 22, 97, 183

Jahn-Graef-Younes method, 194

Kaliszewski problem, 65

Lagrange function, 47
Lagrange multiplier, 47
leader, 184
lower level problem, 184
lower level variable, 185

mathematical program with vector optimization constraints, 184
minimal point, 6
  Edgeworth-Pareto-, 6
  EP-, 6
  ε-EP-, 14, 193
  K-, 6
  local, 14

locally weakly, 14
proper, 12
weakly, 10
weakly EP-, 10
minimal value function, 67
  local, 68
modified Polak method, 55

Newton's method, 111, 128, 130
non-dominated, 6
normal boundary intersection (NBI) method, 41, 53, 130

objective function, 3
optimistic approach, 185
optimization problem
  bicriteria, 6, 17, 168
  biobjective, 6
  tricriteria, 40, 155, 176, 199
ordering
  componentwise, 5
  Löwner partial, 19, 40, 134
  lexicographical, 7
  natural, 5
  partial, 4, 22, 40, 68, 153, 186
  total, 7
ordering cone, 5
  Löwner, 19

parameter set, 31
Pareto filter, 193
partially ordered linear space, 5
Pascoletti-Serafini problem, 23
  algorithm for, 124
  modified, 44
preference order, 4

reference point, 24, 57, 64
reference problem, 67
relative interior, 25

scalarization, 21
sensitivity function, 97
sequential quadratic programming (SQP) method, 114, 127, 151, 177
surrogate worth trade-off method, 97

trade-off, 82, 97, 168
two-level optimization, 184

uniformity level, 105
upper bounds, 21, 49
upper level problem, 184

upper level variable, 184

vector-valued, 3
voxel, 169

weighted Chebyshev norm, 25, 57, 66
weighted minimax method, 58
weighted p-power method, 64
weighted sum, 59
weighted sum method, 21, 61, 144, 148, 156, 167
   generalized, 59
weights, 22, 24, 57, 144

Printed in the United States
121447LV00003B/289-297/P